MECHATRONIC SYSTEMS DESIGN AND SOLID MATERIALS

Methods and Practices

MECHATRONIC SYSTEMS DESIGN AND SOLID MATERIALS

Methods and Practices

Edited by

Satya Bir Singh, PhD

Prabhat Ranjan, PhD

Alexander V. Vakhrushev, DSc

A. K. Haghi, PhD

AAP | APPLE ACADEMIC PRESS

First edition published 2021

Apple Academic Press Inc.
1265 Goldenrod Circle, NE,
Palm Bay, FL 32905 USA

4164 Lakeshore Road, Burlington,
ON, L7L 1A4 Canada

CRC Press
6000 Broken Sound Parkway NW,
Suite 300, Boca Raton, FL 33487-2742 USA

2 Park Square, Milton Park,
Abingdon, Oxon, OX14 4RN UK

Apple Academic Press exclusively co-publishes with CRC Press, an imprint of Taylor & Francis Group, LLC

Library and Archives Canada Cataloguing in Publication

Title: Mechatronic systems design and solid materials : methods and practices / edited by Satya Bir Singh, PhD, Prabhat Ranjan, PhD, Alexander V. Vakhrushev, DSc, A.K. Haghi, PhD.

Names: Singh, Satya Bir, editor. | Ranjan, Prabhat, (Mechatronics professor), editor. | Vakhrushev, Alexander V., editor. | Haghi, A. K., editor.

Description: First edition. | Includes bibliographical references and index.

Identifiers: Canadiana (print) 20200386131 | Canadiana (ebook) 20200386190 | ISBN 9781771889155 (hardcover) | ISBN 9781003045748 (ebook)

Subjects: LCSH: Mechatronics.

Classification: LCC TJ163.12 .M43 2021 | DDC 621—dc23

Library of Congress Cataloging-in-Publication Data

CIP data on file with US Library of Congress

ISBN: 978-1-77188-915-5 (hbk)
ISBN: 978-1-00304-574-8 (ebk)

About the Editors

Satya Bir Singh, PhD
Professor, Department of Mathematics, Punjabi University, Patiala, India

Satya Bir Singh, PhD, is a Professor of Mathematics at Punjabi University Patiala in India. Prior to this, he worked as an Assistant Professor in Mathematics at the Thapar Institute of Engineering and Technology, Patiala, India. He has published about 125 research papers in journals of national and international repute and has given invited talks at various conferences and workshops. He has also organized several national and international conferences. He has been a coordinator and principal investigator of several schemes funded by the Department of Science and Technology, Government of India, New Delhi; the University Grants Commission, Government of India, New Delhi; and the All India Council for Technical Education, Government of India, New Delhi. He has 21 years of teaching and research experience. His areas of interest include the mechanics of composite materials, optimization techniques, and numerical analysis. He is a life member of various learned bodies.

Prabhat Ranjan, PhD
Assistant Professor, Department of Mechatronics Engineering at Manipal University Jaipur, India

Prabhat Ranjan, PhD, is an Assistant Professor in the Department of Mechatronics Engineering at Manipal University Jaipur. He is the author of Basic Electronics and editor of *Computational Chemistry Methodology in Structural Biology and Materials Sciences*. Dr. Ranjan has published more than 10 research papers in peer-reviewed journals of high repute and dozens of book chapters in high-end research edited books. He has received prestigious the President Award of Manipal University Jaipur, India; a Material Design Scholarship from Imperial College of London, UK; a DAAD Fellowship; and the CFCAM-France Award. Dr. Ranjan has received several grants and also participated in national and international conferences and summer schools. He holds a bachelor of engineering in electronics and communication and a master of technology in instrumentation control system engineering from the Manipal Academy of Higher Education, Manipal, India, as well as a PhD in engineering from Manipal University Jaipur, India.

Alexander V. Vakhrushev, DSc
Professor, M. T. Kalashnikov Izhevsk State Technical University, Izhevsk, Russia; Head, Department of Nanotechnology and Microsystems of Kalashnikov Izhevsk State Technical University, Russia

Alexander V. Vakhrushev, DSc, is a Professor at the M.T. Kalashnikov Izhevsk State Technical University in Izhevsk, Russia, where he teaches theory, calculating, and design of nano- and microsystems. He is also the Chief Researcher of the Department of Information-Measuring Systems of the Institute of Mechanics of the Ural Branch of the Russian Academy of Sciences and Head of the Department of Nanotechnology and Microsystems of Kalashnikov Izhevsk State Technical University. He is a Corresponding Member of the Russian Engineering Academy. He has over 400 publications to his name, including monographs, articles, reports, reviews, and patents. He has received several awards, including an Academician A. F. Sidorov Prize from the Ural Division of the Russian Academy of Sciences for significant contribution to the creation of the theoretical fundamentals of physical processes taking place in multi-level nanosystems and the designation of Honorable Scientist of the Udmurt Republic. He is currently a member of the editorial boards of several journals, including *Computational Continuum Mechanics, Chemical Physics and Mesoscopia,* and *Nanobuild.* His research interests include multiscale mathematical modeling of physical-chemical processes into the nano-hetero systems at nano-, micro- and macro-levels; static and dynamic interaction of nanoelements; and basic laws relating the structure and macro characteristics of nano-hetero structures.

A. K. Haghi, PhD

Professor Emeritus of Engineering Sciences, Former Editor-in-Chief, International Journal of Chemoinformatics and Chemical Engineering and Polymers Research Journal; Member, Canadian Research and Development Center of Sciences and Culture

A. K. Haghi, PhD, is the author and editor of 165 books, as well as 1000 published papers in various journals and conference proceedings. Dr. Haghi has received several grants, consulted for a number of major corporations, and is a frequent speaker to national and international audiences. Since 1983, he has served as a professor at several universities. He is former Editor-in-Chief of the *International Journal of Chemoinformatics and Chemical Engineering* and *Polymers Research Journal* and is on the editorial boards of many international journals. He is also a member of the Canadian Research and Development Center of Sciences and Cultures (CRDCSC), Montreal, Quebec, Canada. He holds a BSc in urban and environmental engineering from the University of North Carolina (USA), an MSc in mechanical engineering from North Carolina A&T State University (USA), a DEA in applied mechanics, acoustics, and materials from the Université de Technologie de Compiègne (France), and a PhD in engineering sciences from Université de Franche-Comté (France).

Contents

Contributors

Bhavik A. Ardeshana
Mechatronics Engineering Department, G. H. Patel College of Engineering & Technology, Vallabh Vidyanagar, Gujarat, India, E-mail: ardeshanabhavik@gmail.com

Himanshu Chaudhary
Malaviya National Institute of Technology Jaipur, Rajasthan – 302017, India

Swarup S. Deshmukh
Research Scholar, Department of Mechanical Engineering, National Institute of Technology Durgapur, Durgapur, West Bengal – 713209, India, E-mail: dss.19me1106@phd.nitdgp.ac.in

V. Dhinakaran
Center for Applied Research, Chennai Institute of Technology, Chennai, Tamil Nadu, India

A. Yu. Fedotov
Department of Mechanics of Nanostructures, Institute of Mechanics, Udmurt Federal Research Center, Ural Division, Russian Academy of Sciences, Izhevsk, Russia; Department of Nanotechnology and Microsystems, Kalashnikov Izhevsk State Technical University, Izhevsk, Russia

Arjyajyoti Goswami
Assistant Professor, Department of Mechanical Engineering, National Institute of Technology Durgapur, Durgapur, West Bengal – 713209, India, E-mail: arjyajyoti.goswami@me.nitdgp.ac.in

Vijay S. Jadhav
Professor, Mechanical Engineering Department, Government College of Engineering Karad, Karad, Maharashtra – 415124, India, E-mail: ramakant.shrivastava@gcekarad.ac.in

T. Jagadeesha
Department of Mechanical Engineering, NIT Calicut, India, E-mail: jagdishsg@nitc.ac.in

Umang B. Jani
Mechatronics Engineering Department, G. H. Patel College of Engineering & Technology, Vallabh Vidyanagar, Gujarat, India, E-mail: umangjani@gcet.ac.in

Anand Y. Joshi
Mechatronics Engineering Department, G. H. Patel College of Engineering & Technology, Vallabh Vidyanagar, Gujarat, India, E-mail: anandyjoshi@gmail.com

Jitendra Kumar Katiyar
SRM Institute of Science and Technology, Chennai, Tamil Nadu, India

Pancham Kumar
School of Electrical Skills, Bhartiya Skill Development University Jaipur, Rajasthan, India

A. T. Lekontsev
Department of Nanotechnology and Microsystems, Kalashnikov Izhevsk State Technical University, Izhevsk, Russia

T. Pankaj
ICFAI University, Himachal Pradesh, India

Ajay M. Patel
Mechatronics Engineering Department, G. H. Patel College of Engineering & Technology, Vallabh Vidyanagar, Gujarat, India, E-mail: ajaympatel2003@yahoo.com

Karali Patra
Associate Professor, Department of Mechanical Engineering, Indian Institute Technology, Patna – India

N. R. N. V. Gowripathi Rao
Malaviya National Institute of Technology Jaipur, Rajasthan – 302017, India, E-mail: gowripathiraofmpe@gmail.com

Dhananjay Sahu
Senior Research Fellow, Department of Mechanical Engineering, National Institute of Technology, Raipur – 492 010, India

Raj Kumar Sahu
Assistant Professor, Department of Mechanical Engineering, National Institute of Technology, Raipur – 492 010, India, E-mail: raj.mit.mech@gmail.com

Yu. B. Savva
Orel State University named after I.S. Turgenev, Orel, Russia

Ajay Kumar Sharma
College of Technology and Engineering, MPUAT Udaipur, Rajasthan – 313001, India

Susheela Sharma
Department of Basic Science, Bhartiya Skill Development University Jaipur, Rajasthan, India

Manisha Sheoran
Department of Basic Science, Bhartiya Skill Development University Jaipur, Rajasthan, India

Ramakant Shrivastava
Associate Professor, Mechanical Engineering Department, Government College of Engineering Karad, Karad, Maharashtra – 415124, India, E-mail: vijay.jadhav@gcekarad.ac.in

A. S. Sidorenko
Orel State University named after I.S. Turgenev, Orel, Russia; Ghitu Institute of Electronic Engineering and Nanotechnologies, Chisinau, Republica Moldova

S. B. Singh
Department of Mathematics, Punjabi University, Patiala, India

A. G. Temesgen
Department of Mathematics, Punjabi University, Patiala, India

A. V. Vakhrushev
Department of Mechanics of Nanostructures, Institute of Mechanics,
Udmurt Federal Research Center, Ural Division, Russian Academy of Sciences, Izhevsk, Russia; Department of Nanotechnology and Microsystems,
Kalashnikov Izhevsk State Technical University, Izhevsk, Russia,
E-mail: Vakhrushev-a@yandex.ru

Abbreviations

AJM	abrasive jet machining
Al2O3	aluminum oxide
ANOVA	analysis of variance
a-Si	amorphous silicon
B2B	business to business
B2C	business to consumer
CdTe	cadmium telluride
CIGS	copper indium gallium di-selenide
CNCs	carbon nanocones
CNTs	carbon nanotubes
Cr	chromium
CSC	concentrated solar cells
Des	dielectric elastomers
DFT	density functional theory
DSC	dye-sensitized solar cells
DT	dimensional tolerance
DWCNCs	double walled carbon nanocones
Ea	activation energy
EAM	embedded-atom method
ECM	electrochemical machining
EDM	electro discharge machining
F	force
FEM	finite element method
HSTR	high-strength-temperature-resistant
LAMMPS	large-scale atomic/molecular massively parallel simulator
MD	molecular dynamics
MEAM	modified embedded atom method
MLS	manufacturer liability system
Mo	molybdenum
MRAM	magnetoresistive random-access memory
MRR	material removal rate
MSD	mean square deviation
MSM	molecular structural mechanics

NR	natural rubber
PC	powder concentration
PE	polyethylene
PSC	polymer-based solar cell
PTO	power take-off
pwEDM	powder mixed wire electric discharge machining
Ra	roughness
SCB	surface Cauchy-Born
SD	standard deviation
Si3N4	silicon nitride
SiC	silicon carbide
SIC	strain-induced crystallization
SR	surface roughness
SST	stainless steel
SV	servo voltage
SWCNCs	single-walled carbon nanocones
TEM	transmission electron microscopy
TGA	thermogravimetric analysis
TOFF	pulse off time
TON	on time
TWR	tool wear rate
VMD	visual molecular dynamics
wEDM	wire electric discharge machining
WF	wire feed rate
XRD	x-ray diffraction

Preface

Mechatronics is the synergistic integration of mechanical engineering with electronics and intelligent computer control in the design and manufacturing of industrial products and processes. Mechatronics integrates mechanical systems (mechanical elements, components, and machines), electronic systems (microelectronics, sensor, and actuator technology), and information technology. In this manner, mechatronic systems are a complex integration of extremely advanced technological components, which able to perform tasks with high accuracy and flexibility. This volume is devoted to exploiting some advancement in the wide field of mechatronics.

Microporomechanics deals with the mechanics and physics of multiphase porous materials at nano and micro scales. It is composed of a logical and didactic build-up from fundamental concepts to state-of-the-art theories.

This volume provides practicing mechanical/mechatronics engineers and designers, researchers, graduate, and postgraduate students with a knowledge of mechanics focused directly on advanced applications.

Machining of materials has been an important and highly useful process catering to the needs of humans. With the advent of new and novel materials, the conventional approaches to machining are no longer sufficient and the industry is continuously shifting from traditional to nontraditional machining processes. The powder mixed wire electric discharge machining (pwEDM) is the modified version of the wire electric discharge machining (wEDM). In wEDM, the powder is added to the dielectric fluid through mechanical stirring. The outputs obtained in pwEDM are largely dependent on the properties of the powder which is added to the dielectric fluid. The physical, electrical, and thermal properties of powder play an important role in determining the material removal rate (MRR) and surface roughness (Ra) of the machined component.

Chapter 1 is focused on reducing the surface roughness and increasing the material removal rate of the machined specimen, i.e., AISI 4140. In this experimentation, graphite powder of particle size 25 μm was added in a dielectric fluid. The main reasons behind the addition of graphite powder were its high thermal conductivity and low density. Owing to the high thermal conductivity, the graphite powder dissipates more heat from the cutting area which reduces the crater size, as compared to the normal wEDM. This phenomenon helps in improving the surface quality. Also, the low density of the graphite powder ensures that it mixes properly with the dielectric fluid.

The input process parameters considered in this experimentation are pulse on-time (T_{ON}), pulse off-time (T_{OFF}), servo voltage (SV), wire feed (WF), and powder concentration (PC). The response variables that are considered in this experimentation are MRR and Ra. The experiments were conducted as per the Taguchi methodology using the L27 orthogonal array.

It was observed that after the selection of parameters as per Taguchi methodology the surface roughness was reduced by 49.51% and the material removal rate was increased by 73.51%.

Agricultural tillage plays an essential role in the farming community. There are different agricultural operations which contribute to the overall development of the crop, but among them, agricultural tillage plays an important role. Various unit operations are tillage, sowing and fertilizing, irrigation, harvesting, and post-processing activities. Chapter 2 deals to design a four-bar mechanism for a concept known as vibratory tillage, which can be used for agricultural soil manipulation operation with improved soil properties and less power consumption. Mechanism design is through the path generation process, and graphical and analytical procedures are used to design a mechanism for a particular vibratory tillage tool trajectory in soil. Three and four precision methods are used to design a four-bar mechanism through which the tool follows the path very accurately. The mechanism design is validated through MATLAB, and it is confirmed that the tool passes through the selected precision points correctly. Also, it is found that the mechanism dimensions obtained through the graphical and analytical techniques are exact.

In abrasive jet machining, a focused stream of abrasive particles carried by high-pressure air or gas is made to impinge on the work surface through a nozzle, and work material is removed by the erosion of high-velocity abrasive particles. The AJM differs from sandblasting and in that the abrasive is much finer and the process parameters and cutting action are carefully controlled. AJM is mainly used to cut intricate shapes in hard and brittle material which are sensitive to heat and chip easily.

The process is also used for deburring and cleaning operations. AJM is inherently free from chatter and vibration problems. The cutting action is cool because the carrier gas serves as a coolant. The high-velocity stream of abrasives is generated by converting the pressure energy of carrier gas or air to its Kinetic energy and hence high-velocity jet. Abrasive jet machining consists of a gas propulsion system, abrasive feeder, machining chamber, AJM nozzle, and abrasives. Aluminum oxide (Al_2O_3), silicon carbide (SiC), glass beads, crushed glass, and sodium bicarbonate are some of the abrasives used in AJM. The selection of abrasives depends on MRR, type of work material, machining accuracy. This chapter deals with the process details, experimental setup, and process parameters used in AJM. The mechanics of metal removal in AJM is presented in Chapter 3. Mathematics of material removal models for ductile and brittle materials are discussed. Finally, practical examples are given to enhance the understanding of the mechanics of material removal in AJM.

In Chapter 4, an approach to modeling and simulation of single-wall carbon nanocones (SWCNC) has been suggested for mass sensing applications. Finite element modeling and dynamic analysis of SWCNC with cantilever beam boundary condition, various disclination angles of 60°, 120°, 180°, 240°, and 300° and 10, 15, 20 A° lengths have been completed using atomistic molecular structure. This study has been conducted to evaluate and identify the difference in fundamental frequencies shown by these nanodevices when subjected to sensing applications. The study also displays the outcome of alteration in the length of nanocones on the vibrational frequencies. It is witnessed that increasing length of a SWCNC with the same apex angle outcome in a drop in the fundamental frequency. Additionally, it is clear from the outcomes that SWCNC with greater apex angles displays lesser values of fundamental frequencies. Original and defective single-walled nanocones have been analyzed to study the effect of defects like vacancy defect and stone wales defect. The results show the fact that with the change in the disclination angle and defects, there is a significant amount of variation in the stiffness due to the different positions on defects in nanocones. The outcomes propose that smaller lengths of nanocones are good contenders for sensing applications as they display extensive variation in the fundamental frequencies. It also shows that the mass increases a certain limit.

Ultrasonic machining is a nontraditional process, in which abrasives contained in the slurry are driven against the work by a tool oscillating at low amplitude (25–100 microns) and high frequency (15–30 kHz). It is employed to machine hard and brittle materials (both electrically conductive and non-conductive material) having hardness usually greater than 40 HRC. The process was first developed in the 1950s and was originally used for finishing EDM surfaces. In this chapter, a detailed process and its process parameters are discussed. A brief summary of the equipments and tool configurations is presented. The materials removal model for both ductile and brittle materials is discussed in great detail. Various velocity transformers design aspects and criteria are discussed in Chapter 5 with practical examples.

Chapter 6 contributes significant information that encourages the study of pre-strain induced adaptation in chain network hence, material behavior for improved actuation strain. Controlled for a certain level, the stiffening of chains in VHB elastomer is found to turns into crystallization, provided desire temperature, and strain rate.

Since two decades, pre-strain in dielectric elastomers is persisted as a proficient technique to improve the performance of soft actuators, sensors, and energy harvesters configuring them. But, information on the perception of pre-strain induced molecular re-arrangement and crystallization in the dielectric elastomer is yet to be established. Studies of these structural modifications are important to consider because the change in electromechanical behavior is frequently attributed to chain entanglement for applied pre-strain. Here, the effects of pre-strain on the behavior of VHB 4910 dielectric elastomer are deliberated towards electromechanical devices. The influence of pre-strain on electromechanical properties is conferred relating to the phenomenon of macromolecules and chain orientation. The chain entanglement is found to cause a stiffening effect and lead strain-induced crystallization. So, the required strain percentage, activation energy, and time of crystallization are estimated based on experimental and analytical work involving the strain rate. These results evidence the kinetics of crystallization in the mentioned elastomer with empirical equivalence, and is an inventive contribution towards complete understanding on the effect of pre-strain for actuator application.

Increasing changes in weather conditions like up-surged global temperature, pollution, population explosion, and economic backwardness are the major problems faced by our society in the present century. The natural resources have been exploited to their peek by the increasing population. Emerging industries and enhancement of the lifestyle of people across the globe have further exhausted the conventional sources of energy with mother Earth being in its usual form. The growing taste of urbanization and modernization in developing countries are causing major repercussions for the environment and energy sector. The requirement of energy and its associated amenities indulges humankind in the blooming of social, economic, and health benefits. To meet these requirements, the necessity of renewable energy sources was emphasized. With the ubiquitous presence of the Sun, solar energy proved to be an inexpensive and versatile source of energy. Further, the provision of government subsidies the solar photovoltaic power bloomed across the world. Using the life cycle assessment (LCA) approach, an in-depth study of the various inputs and outputs in a solar PV system is studied. Though with increasing agitation in climate perturbed a sustainable and greener route to decrease the greenhouse gaseous emissions is mandatory. The use of renewable sources of energy in a clever mode will bring drastic changes in the improvement of energy issues in a greener way. The clean and green way of renewable energy resources has to be applied in every niche of the world. Solar energy proved to be a major contributor to this task. The governmental initiative of more PV installations is leading to the macroscopic proliferation of solar photovoltaic waste accumulation in the country, which has become the utmost issue of the hour to be handled by PV scarp management and recycling policy. Existing PV technologies will be evaluated and their effects on the environment, human health social, and economic aspects will be analyzed in depth. Major PV module recycling methods are evaluated, and the material recovery in economic terms is recorded. With the sky rocking PV scrap, India is obligate for inclination towards the PV recycling policy framework. The main focus of Chapter 7 is towards the effective remedy of the environmental and socio-economic impact arising in the life cycle assessment of c-Si, CdTe, CdS, and CIGS. A sustainable policy is to be put forward to potentially tackle the upcoming problems from the PV waste generation and accumulation.

Chapter 8 is devoted to the study of the formation processes and analysis of the structure of a superconducting spin valve based on a multilayer superconductor-ferromagnet nanostructure. The relevance of the research is due to the need to develop an energy-efficient element base for microelectronics based on new physical principles and the advent of devices based on spin and quantum-mechanical effects. The superconducting spin valve being developed is a multilayer structure consisting of ferromagnetic cobalt nanofilms, which are separated by niobium superconductors. The studies were carried out using molecular dynamics modeling. As the interaction potential of atoms in the simulated system, the modified immersed atom method is used. The spin valve was formed by layer-by-layer deposition of elements in a vacuum. The atom deposition process was simulated in a stationary temperature regime. The chapter presents a simulation of the deposition of the first few layers of a nanosystem. The atomic structure of individual nanolayers of the system is considered. Particular attention is paid to the analysis of the atomic structure of contact areas at the junction of the layers, since the quality of the layer interface plays a crucial role in creating a workable device. Three temperature deposition regimes were implemented: 300, 500, and 800 K. Calculations showed that with an increase in temperature there is a rearrangement of the structure of the system layers and their loosening. The structure of the nanolayer from niobium is close to crystalline with division into regions of different crystallographic orientations of atomic layers. For cobalt nanofilms, an amorphous structure is more characteristic. The obtained simulation, results can be used in the development, and as well as optimization of technologies for the formation of spin valves and other functional elements for spintronics.

Chapter 9 formulates the problem of deformation and fracture of nanocomposites by the molecular dynamics method. To describe the interatomic potential used embedded atom. Molecular dynamic modeling of uniaxial tension of a layered Al/Cu nanocomposite has been performed. Deformation nanocomposite is carried by elastic and plastic deformation of the material to fracture. The basic parameters of the deformation of materials are investigated-deformation, stress, temperature, and atomic structure of the material. Modeling showed that when the stresses in the sample reached the elastic limit, nucleation of defects in the crystal lattice of the material and their propagation through the crystal in the form of shifts and rotations of atoms in the crystal planes were observed. The areas of nucleation of plastic strains and the formation of defects are determined. The maximum destruction of the material occurred at the interface of the components of the nanocomposite.

The objective of Chapter 10 is to derive the problem of elastoplastic modeling of an orthotropic boron-aluminum fiber-reinforced composite thick-walled rotating cylinder subjected to a temperature gradient by using Seth's transition and generalized strain measure theory. The combined effects of temperature and angular speed have been presented numerically and graphically. Seth's transition theory does not require the assumptions: the yield criterion, the incompressibility conditions, the deformation is small, etc., and thus solves a more general problem. This theory utilizes the concept of generalized strain measure and asymptotic solution at the turning points of the differential equations defining the deformed field. It is seen that cylinders having smaller radii ratios require higher angular speed for yielding as compared to cylinders having higher radii ratios. With the inclusion of thermal effects, the angular speed increased for initial yielding to a smaller radii ratio, but for the fully plastic state, the angular speed is the same. It is observed that the maximum circumferential stress occurs at the internal surface for both transitional and fully plastic state at any temperature and angular speed.

CHAPTER 1

Parametric Study and Optimization of Parameters in Powder Mixed Wire-EDM Using Taguchi Analysis

SWARUP S. DESHMUKH,[1] ARJYAJYOTI GOSWAMI,[2] RAMAKANT SHRIVASTAVA,[3] and VIJAY S. JADHAV[4]

[1]*Research Scholar, Department of Mechanical Engineering, National Institute of Technology Durgapur, Durgapur, West Bengal – 713209, India, E-mail: dss.19me1106@phd.nitdgp.ac.in*

[2]*Assistant Professor, Department of Mechanical Engineering, National Institute of Technology Durgapur, Durgapur, West Bengal – 713209, India, E-mail: arjyajyoti.goswami@me.nitdgp.ac.in*

[3]*Associate Professor, Mechanical Engineering Department, Government College of Engineering Karad, Karad, Maharashtra – 415124, India, E-mail: vijay.jadhav@gcekarad.ac.in*

[4]*Professor, Mechanical Engineering Department, Government College of Engineering Karad, Karad, Maharashtra – 415124, India, E-mail: ramakant.shrivastava@gcekarad.ac.in*

ABSTRACT

Machining of materials has been an important and highly useful process catering to the needs of humans. With the advent of new and novel materials, the conventional approaches to machining are no longer sufficient and the industry is continuously shifting from traditional to non traditional machining processes. The powder mixed wire electric discharge machining (pwEDM) is the modified version of the wire electric discharge machining (wEDM). In wEDM, powder is added into the dielectric fluid through mechanical stirring. The outputs obtained in pwEDM are largely dependent on the properties of the powder, which is added to the dielectric fluid. The physical, electrical, and thermal properties of powder play an important role in determining the material removal rate (MRR) and surface roughness (Ra) of the machined component.

This study is focused on reducing the surface roughness and increasing the material removal rate of the machined specimen (i.e., AISI 4140). In this experimentation graphite powder of particle size, 25 μm was added in the dielectric fluid. The main reasons behind the addition of graphite powder were its high thermal conductivity and low density. Owing to the high thermal conductivity, the graphite powder dissipates more heat from the cutting area, which reduces the crater size, as compared to the normal w-EDM. This phenomenon helps in improving the surface quality. Also, the low density of the graphite powder ensures that it mixes properly with the dielectric fluid.

The input process parameters considered in this experimentation are pulse on time (T_{ON}), pulse off time (T_{OFF}), servo voltage (SV), wire feed (WF), and powder concentration (PC). The response variables that are considered in this experimentation are material removal rate (MRR), and surface roughness (Ra). The experiments were conducted as per the Taguchi methodology using the L27 orthogonal array.

It was observed that after the selection of parameters as per Taguchi methodology the surface roughness reduced by 49.51% and the material removal rate increased by 73.51%.

1.1 INTRODUCTION

Machining can be defined as the process of selective removal of work material to impart the desired shape and hence a desired functionality to the work material. Machining is a very useful process, which has been frequently utilized throughout the history of humans. Arguably, the earliest example of machining was when the primitive human picked up a stone and selectively chiseled away the edges to generate what was the first tool in the history of human civilization [1]. Soon our ancestors realized that by selective removal of material from a chosen substrate one can develop a particular tool for a specific purpose, which can reduce the effort and increase the benefit of a process.

With the progress of civilization, the demand and needs of human beings also increased. In order to fulfill the needs, the work material and the machining process also improved and increasingly became more capable as well as efficient. With the invention of the steam engine (1775), civilization entered a new age. The output of the machines as well as the processing capabilities increased many folds as compared to earlier machining methods. At this time the most commonly used material in the industry was steel (different variants) and the tool used for selective removal of the work material was carbide. But, as technology progressed, different materials were chosen for generating specific components, which were to be used in highly specialized applications. Different industries like the automobile industry, shipbuilding, aviation, etc., started using different grades of materials.

The use of different materials combined with an increase in demand (owing to World War I and World War II) led to the next stage in the industrial revolution wherein CNC machines were developed and implemented. Different CNC machine tools like CNC lathe, CNC milling, CNC drilling, etc., were developed. Apart from being highly flexible in terms of the type of output that can be generated through these machines, CNC machines were also capable of processing a wide variety of work material through the use of different cutting tools. Also, the rate of output from CNC machines was much higher compared to conventional machining methods.

Despite rapid technological advancements in the field of conventional machining processes, the machining of carbides and other hard-to-machine materials has been limited to the diamond wheel grinding for a long time. Diamond grinding is not a cost-effective process since diamond grits are not readily or cheaply available. Also, the industries of aviation, automobile, and marine made very rapid progress, which resulted in them using different alloys of materials. Such high-strength-temperature-resistant (HSTR) alloys cannot be processed through conventional means. The required shapes and sizes of the machined component are becoming increasingly complex such that conventional methods of machining are either too time-consuming or are entirely incapable of generating such parts. Thus, neither the conventional techniques of machining are in a position to meet the challenges posed by the new development of materials nor is there any greater scope for further development. Therefore, it is clear that some new strategies of machining must be developed in order to deal with the challenges created by the development and use of the hard-to-machine and high strength temperature resistant alloys.

The conventional machining process of selectively removing work material through chips is not satisfactory, economical, and sometimes even impossible for the following reasons:

1. The material possessing high hardness, strength, and brittleness properties will not be economically cut by conventional machining processes.
2. The workpiece is too flexible, slender, or delicate to withstand the cutting or grinding forces, or the parts are difficult to fix or clamp in work holding devices.
3. The shape of the part includes such features as internal and external profiles or small diameter holes.
4. Surface finish and DT (dimensional tolerance) requirements are more rigorous than those obtained by other processes.

The major types of nontraditional machining processes and their characteristics have been listed in Table 1.1. The classification has been done on the basis of the principle of operation and the characteristics of the machined component.

TABLE 1.1 Characteristics of the Nontraditional Machining Process

Process	Characteristics
Chemical machining	Shallow removal on flat surfaces suitable for low production runs.
Electrochemical machining (ECM)	Ion dissolution phenomena used for MRR, coulombs law applicable here, contour shape deep cavities developed.
Electric discharge machining (EDM)	Spark energy is used for eroded material from a workpiece. Hard materials are easily machined.
Wire EDM	The advanced version of EDM, electrode in the form of wire, is used for contour and complex shapes cut from a conductive material.
Electron beam machining	The high velocity of electrons from an electron gun are bombarded on W/P material K.E. converts to thermal energy and develops heat used for melting. Small holes and cutting slots easily machined.

TABLE 1.1 *(Continued)*

Process	Characteristics
Laser beam machining (LBM)	This process is carried out in a vacuum; the laser gun is used for the developing laser. Small holes can be easy with the help of LBM.
Water jet machining (WJM)	High-pressure water energy is used for cutting of soft and brittle material up to 25 mm thickness.
Abrasive water jet machining (AWJM)	It is the further advancement of water jet machining. Abrasive particles are added in water used for cutting contour shape from the brittle and non-conductive workpiece.

1.1.1 *WIRE ELECTRIC DISCHARGE MACHINING (wEDM)*

From the different nontraditional machining processes listed above, wire electric discharge machining (wEDM) is particularly suited for cutting different types of conductive materials irrespective of their hardness or strength. wEDM is basically a modified version of electric discharge machining (EDM). The electrode used in the wEDM is in the form of wire and its diameter is varied from the 0.18 to 0.30 mm. The material used as the wire is brass, zinc-coated brass, diffused brass, molybdenum, etc.

The wire is continuously fed from a spool and is guided by a lower and upper nozzle. Similar to the EDM, the potential difference needs to be applied between work material (anode) and the wire (cathode). A small gap is maintained between the wire and the work material. A dielectric fluid, either kerosene or deionized water, is present which surrounds the workpiece and the wire.

Upon application of potential, the electrons move from the cathode towards the anode and in the process knocks out electrons from the dielectric fluid, thus generating even more number of electrons. These electrons move towards the anode (work material) creating a conductive channel between them. This results in a spark between the anode and the cathode. The spark releases energy in the form of thermal energy at the work material and the temperature rises to ~10,000°C which results in melting and then evaporation of the work material thus achieving material removal. During machining small debris is generated due to material removal, which is flushed through the dielectric fluid. Due to the spark energy, the wire is continuously eroded and is used only once for machining operation. As mentioned earlier wire is fed continuously from the spool, this ensures that a new portion of the wire is continuously available in the machining area. It helps in improving the product reliability. The wire travels at a constant velocity in the range of 1 m/min to 10.0 m/min, and a constant gap (kerf) is maintained during the cut.

In wEDM, the wire travels according to the programmed path given by CNC in slow motion. Normally a 5-axis machine is used for wEDM. Normal cutting motion takes place in the X and Y axis and taper cutting operation takes place in the U and V axis. The movement in case of taper turning is controlled through the joystick. The upper guide moves in the Z-axis, that is, vertically up or down. The working principle of wEDM is shown in Figure 1.1.

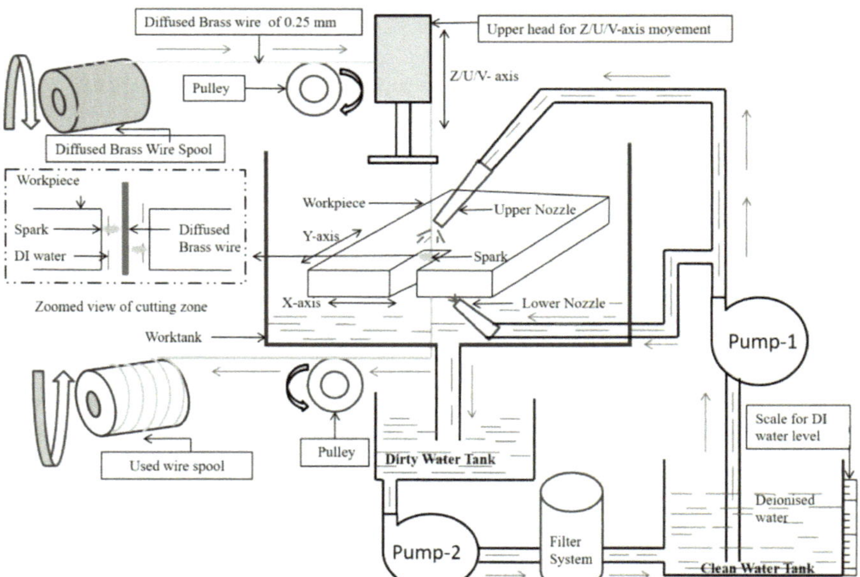

FIGURE 1.1 Schematic of wire electric discharge machining.

wEDM has many benefits, which are listed below:

- **Greater Manufacturing Capabilities:** wEDM can cut the workpiece very precisely irrespective of the hardness of the workpiece. It can cut any complex contours as per requirement. Thermal damages during machining are very less when compared to laser beam machining.

- **Production Reliability:** wEDM uses wire as a tool. It is ensured that during machining the wire is exposed to the work material only once. This ensures that the last part cut from the workpiece is similar to the first part in all aspects.
- **Developed Burr and Stress-Free Machined Part:** In wEDM, the tool (wire) does not come into contact with the workpiece. The wire electrode travels in the programmed path and cutting takes place due to spark energy so there is no chance for the development of burr on the surface and the machined surface is free of sharp edges. Due to this, there are no internal stresses in the workpiece.
- **Tight Tolerances and Excellent Finishes:** The wire travels according to a programmed path developed CNC and the control cabinet in the machine controls the movement of the wire. Thus, there are no chances of any manual error in the process. wEDM manufactures part having high accuracy near about 0.0025 mm and results in a good surface finish.
- **Reduced Costs:** As discussed, wEDM being a non-contact operation does not induce any stress in the workpiece or the tool. This ensures that the machined parts have more life as compared to parts machined by conventional processes (due to residual internal stresses in machined parts, it can be warped at a later stage when the stresses are released). Also, since very tight tolerances and complicated profiles can be realized with the wEDM process, the final product generally does not require any post-processing operations like grinding. This reduces the overall cost of the final product made through wEDM.

1.1.2 POWDER-ASSISTED WIRE ELECTRIC DISCHARGE MACHINING (pwEDM) PROCESS

It was theorized that the presence of powder in the spark zone could alter the process of wEDM. The best way to introduce powder in the spark region is to mix it with the dielectric fluid. The response obtained after mixing the powder in the dielectric fluid is dependent on the physical, electrical, and thermal properties of the powder. For example, a powder particle having low density and high thermal conductivity mixes properly with the dielectric fluid (owing to low density) and dissipates more heat (owing to high thermal conductivity) which is generated at the cutting area. This results in better surface finish. Conversely, a powder particle having lesser thermal conductivity results in a concentration of heat at the point of spark, which results in greater MRR but poor surface finish. The powder is mixed with the dielectric fluid in a separate tank (with the help of a stirrer) and pumped up to the machining zone through upper and lower nozzles. The schematic of powder-assisted electric discharge machining is shown in Figure 1.2.

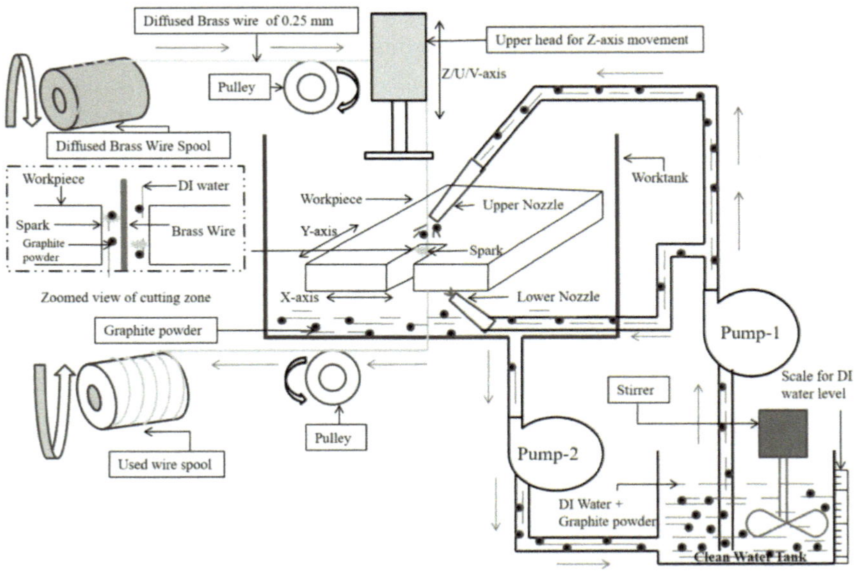

FIGURE 1.2 Schematic of powder-assisted electric discharge machining (pwEDM) (dotted block shows the schematic of close up view of the machining zone).

1.1.3 TAGUCHI METHOD

Taguchi method uses an orthogonal array tool, which provides a combination of process parameters. It helps in the collection of data in experimentation and carrying out the analysis for finding out the most influencing parameter from the process. Taguchi method reduces the number of experimental trials to be conducted and thus saves time, money, and effort.

Two major tools used in the Taguchi method are:

1. **Orthogonal Array:** The orthogonal array is a set of the matrix, which helps to reduce the total number of experiments, indirectly it helps in making the process robust. The selection of orthogonal array is dependent on the degree of freedom of the process and orthogonal array must be greater than or equal to the degree of freedom of process [8].

$$\text{Degree of freedom of factor} = \text{Number of levels} - 1 \quad (1.1)$$

$$\text{Degree of freedom of process} = \sum \text{Degree of freedom of factors} \quad (1.2)$$

Taguchi represents an orthogonal array as:

$$L_N(S^K) \quad (1.3)$$

where, S = number of levels for each factor; k = maximum number of factors whose effects can be estimated without any interaction; N = total number of trials during experimentation.

Example: $L_9 = (3^4) - 9$ experimental runs, 4 factors with 3 levels.

2. **Signal to Noise Ratio:** The variation between the experimental value and desired value can be calculated with the help of the Taguchi quality loss function. The loss function is converted into a utility function. The utility function is also known as the signal to noise ratio. The signal represents a desirable quantity and noise is the undesirable quantity, a higher value of the signal to noise ratio is selected because it contains a lower noise value. Signal to noise ratio has three types:

   i. Lower the better;
   ii. Nominal the best, and
   iii. Higher the best.

In the case of powder mixed wire electric discharge machining (pwEDM) the response variable such as material removal rate is required to be maximum because of that, it undergoes third category, that is, higher the best. The surface roughness of the workpiece goes in lower the better type category.

Signal to noise ratio is can be calculated as follow:

$$\text{S/N ratio} = -10 \log (L_{ij}) \tag{1.4}$$

where, L_{ij} = Loss function.

The loss function for lower the better type and higher the best type can be calculated as follows:

$$\text{For lower the better type, } L_{ij} = \left[\frac{1}{n}\sum_{i=1}^{n} y_i^{2}\right] \tag{1.5}$$

$$\text{For higher the best type, } L_{ij} = \left[\frac{1}{n}\sum_{i=1}^{n} \frac{1}{y_i^{2}}\right] \tag{1.6}$$

1.2 PREVIOUS WORK AND SCOPE OF PRESENT WORK

1.2.1 LITERATURE REVIEW

With the advent of new materials and requirements from the output, new and novel techniques have been developed to fulfill those requirements. The field of EDM has progressed due to the challenges faced by the modern manufacturing industries, from the development of new materials that are hard and difficult-to-machine such as tool steels, composites, ceramics, superalloys, hastelloy, nitralloy, heat resistant steel, etc. [2]. EDM also finds applications in different areas such as fabricating medical and surgical instruments, optical, dental, and sports, etc., [3].

Studies have been conducted on EDM of a variety of materials. EDM has been found to be a useful tool for the machining of hardened tool steel using electrodes of copper, brass, tungsten, and aluminum [4]. The complete study involved only the discharge current and studied its effect on the various output parameters like material removal rate, surface roughness, diametral overcut, and electrode wear. It was observed that the copper electrode provided the highest material removal rate but the tungsten electrode had a minimum tool wear rate.

Distilled water and kerosene are most commonly used as dielectric fluids during EDM or any of its variants. This was established through experimental work carried out on the EDM of Ti-6Al-4V [5]. It was observed that the material removal rate is higher when distilled water is used as compared to kerosene. This was attributed to the formation of TiC when kerosene is used. The formation of carbides necessitates the use of higher discharge energy and thus causes further retardation of the machining process.

The influence of EDM process parameters has been studied during the machining of X200Cr15 and 50CrV4 [6]. The parameters that were considered for the study were discharge current and pulse discharge energy. It was observed that an increase in discharge energy resulted in instability of the process and the surface became increasingly rough.

EDM has some limitations due to which it cannot be used in tight corners or to fabricate features of smaller dimensions. While EDM is used for generating blind holes on a substrate, wEDM is suitable for through cutting of work materials in any required shape. For this reason, wEDM is a very highly utilized process in the machining industry. Another major difference between EDM and wEDM is that the tool in wEDM is very thin as compared to the electrode in EDM. The thin tool (which is actually a wire) enables machining over very complicated or tight corners. Studies have been conducted to find out the influence of different process parameters on the output such as material removal rate, surface roughness, etc., for wEDM.

Taguchi L27 orthogonal array was used for the optimization of input parameters (pulse on time, pulse off time, current, bed speed) and output parameters (MRR, dimensional error, electrode wear, and surface roughness) [7]. Molybdenum wire of 0.18 mm diameter was used as an electrode. It was concluded that pulse on time has more effect on surface roughness while current has a greater effect on accuracy and MRR of the output.

Multi-objective optimization of input parameters like a pulse on time, pulse off time, peak current, and wire offset was carried out for simultaneously getting maximum MRR and less surface finish and wire wear ratio [8]. Brass wire of 0.25 mm diameter was used as an electrode for the study. It was observed that pulse on time is a major influencing factor for MRR (52.31%) and surface roughness (74.69%), and the wire wear ratio is mainly affected by the wire-offset parameter. Sample machined at high-energy input condition gives a rough surface finish with a lot of built-up edges and better surface finish is obtained by low energy input condition.

wEDM has also been used for surface modification of tungsten carbide-cobalt (WC-5.3%Co.) alloy using aluminum and silicon powder of 400-mesh size [9]. The Si powder is added in the electrolytic passes between the workpiece and electrode. It results in chain formation between the wire electrode and workpiece, also known as the bridging effect. Due to the bridging effect, the insulting strength of electrolyte is decreased while its conductivity is increased. This increases the sparking rate, which ultimately increases the material removal rate.

From the studies conducted by various researchers, it was observed that adding a conductive powder in the EDM process could influence the output obtained from the EDM. The nature of influence (positive or negative) depends upon the properties of the powder [10]. Various works have been carried out in powder-assisted EDM. Powder-assisted EDM has been carried out for machining of die steels [11], titanium alloys [12], metal-matrix composites [13], inconel [14], etc. Different modeling and optimization techniques have also been implemented for selecting the process parameters like the Taguchi utility method [15], response surface methodology [16], etc. But very little work has been carried out for pwEDM.

pwEDM has been used for surface modification of WC-Co alloy [17]. The powder has been added to the machining zone by mixing it with the dielectric fluid using a mechanical stirrer. The different process parameters like peak current, pulse on-time, pulse off-time, and servo-voltage were investigated to find their influence on material transfer, crack formation, etc. From the study, it was observed that Silicon powder yielded better results as compared to aluminum powder.

pwEDM has been modified through the use of nanopowder for machining of gold-coated silicon [18]. The effect of different nanopowder concentrations on the material removal rate of silicon was studied and it was observed that the introduction of powder into the machining zone increases the material removal rate of the work material.

1.2.2 SCOPE OF PRESENT WORK

As seen from the previous discussions, we can conclude that pulse on-time– T_{ON} (μs), pulse off-time – T_{OFF} (μs), servo voltage – SV(volt), wire feed rate– WF (m/min), and powder conc. (g/l) are the important process parameters which need to be controlled in order to achieve optimum performance in terms of material removal rate (MRR), and surface roughness (SR).

The aim of this work is to control the input process parameters of pwEDM such as T_{ON}, T_{OFF}, SV, WF, PC to control the output such as surface roughness (SR) and material removal rate (MRR).

1.2.2.1 PULSE ON TIME (T_{ON})

It is generally represented as T_{ON} and expressed in microseconds (μs). It represents the time duration during which the current is applied between work material and the wire. At a high value of T_{ON} large amount of current is applied between the workpiece material and tool electrode for a longer duration, due to which a large amount of heat energy is produced and this energy results in a higher material removal rate. Similarly, at a low value of T_{ON} less amount of material is removed from the workpiece.

1.2.2.2 PULSE OFF TIME (T_{OFF})

It is generally represented as T_{OFF} and expressed in microseconds (μs). During this time current is absent between the workpiece and tool electrode due to this reason no energy is present between the gap. In this time cutting area is cooled and the debris particle produced after machining is flushed. A low value of pulse off time means more time is given to the spark energy because of that more material is removed from the workpiece, which is helpful in increasing the material removal rate.

1.2.2.3 SERVO VOLTAGE (SV)

Servo voltage is the set voltage at the spark gap or reference voltage. Servo voltage controls the forward and backward movement of the wire head. If the mean SV is greater than the reference voltage, the forward motion of the wire takes place. Similarly, if applied SV is lower than the mean SV it results in the backward movement of the wire. A small value of reference or servo voltage, the actual gap between the workpiece and wire electrode becomes smaller because of which more number of electric discharges take place causing increased material removal rate.

1.2.2.4 WIRE FEED RATE (WF)

The main reason behind the high machining cost through pwEDM is the high cost of the wire electrode. The wire electrode amounts to approximately 70–75% of the machining cost. It is beneficial to set the wire feed rate at minimum value if the wire breakage does not occur. The wire feed rate is generally expressed in m/min. At high cutting speed, if the wire feed rate is low it will result in repeated breakage of wire. Because of that, it is necessary to set the wire feed rate at an appropriate value.

1.2.2.5 POWDER CONCENTRATION (PC)

Powder causes improvement in MRR and SR. It makes a bridge between tool and workpiece so it is very important to check the effect of powder by varying the electrical parameters during machining so that the combined effect of powder and parameters will give the best possible output. There are many optimization methods so proper selection of optimization methods is also important for good results in less time. The important response variables in pwEDM material removal rate, and surface roughness.

1.2.2.6 MATERIAL REMOVAL RATE (MRR)

The material removal rate (MRR) is the amount of material removed per minute from the workpiece during machining. The principle of material removal has already been explained in the previous section. After machining small debris particles are produced which are flushed out with the help of deionized water. MRR is influenced by the thickness of material used for machining, properties of the workpiece, type of wire electrode, and in the case of pwEDM properties of powder added in the electrolytic solution.

1.2.2.7 SURFACE ROUGHNESS (RA)

Surface roughness is measured with the help of a surface roughness tester. The probe of surface roughness tester is moved perpendicular to the lay direction. There are three methods for surface roughness measurement. First, is the average peak to valley height, second is the root mean square method, and third is the centerline average value method. Mostly, the centerline average value method is used. Ra is generally expressed in the micron (μm).

1.3 EXPERIMENTAL SETUP AND METHODOLOGY

The study conducted on the pwEDM process required several components. Major among which were the pwEDM machine, the tool (wire) used, selection of work material and powder for the work, the process parameters used for the experiments, and the instrumentations used for carrying out the measurements.

1.3.1 THE MACHINE

Machining has been done on Electronica Ecocut CNC WEDM machine Model-ELPULS-15, manufactured by Electronica Machine Tools Limited, Pune, India. This machine moves in five axes, that is, X-Y-Z-U-V. For normal cutting operation, X and Y-axes are used and for taper cutting action, U and V axes are utilized. The wire is continuously fed from the spool and guided by the lower and upper nozzle. The used wire is collected in a separate tank. Figure 1.3 shows the wEDM machine used for experimentation. Deionized water is present in a clean tank from where it passes to the resin tank with the help of pump 1. The debris particle produced during machining is flushed out with the help of deionized water and this deionized water passes through the dirty tank. After that, it passes through a filter and finally reaches a clean tank. The set up used for carrying out the experimentation is shown in Figure 1.3.

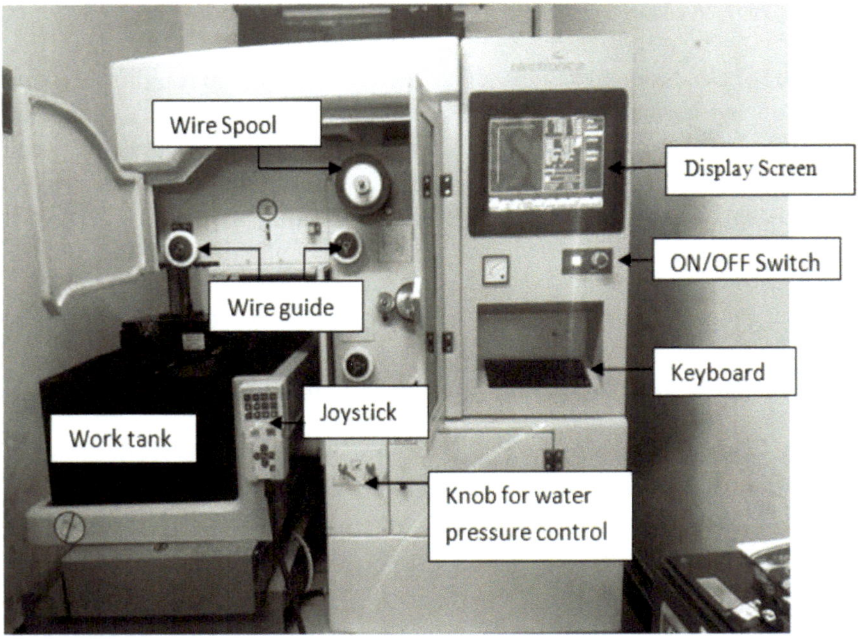

FIGURE 1.3 Set up for the pwEDM.

For pwEDM, the setup was modified. In this setup, powder mixed deionized water from the clean tank passes through pump 1, bypasses the resin tank, and directly goes to the area where pwEDM takes place. After the addition of powder in a clean tank, its conductivity is to be checked by a conductivity meter. The used deionized water again passes through a clean tank and the water is recirculated until the first run of the experiment is completed. After completing the first run powder, mixed deionized water in a clean tank is flushed out by opening the outlet valve before starting the second run of experimentation. Then the clean tank is cleaned with the help of a lintless cloth. The fresh deionized water is taken from the deionized water plant and added to the clean tank. Thereafter, graphite powder is added to the clean deionized water. A mechanical stirrer is attached to the clean tank for continuous stirring action. This helps in avoiding sedimentation and ensures proper mixing of powder in a dielectric fluid.

1.3.2 THE WIRE (TOOL) USED

The ideal wire electrode possesses good electrical conductivity, good tensile strength, and good flushing properties. The diameter of the wire generally varies from 0.18 to 0.30 mm. For the experimentation, a diffused brass wire of 0.25 mm diameter was used.

1.3.3 SELECTION OF WORK MATERIAL

For the present study, AISI 4040 was selected. AISI 4040 is quite commonly used as a material for the fabrication of dies and it is also used extensively in the defense sector. It requires precise machining of the material with the good surface finish as well so that the post-processing costs can be reduced. AISI 4040 has a very high hardness hence pwEDM is a suitable process for machining this material. AISI 4140 is also known as chromium (Cr)-molybdenum (Mo) alloy steel, this grade is mostly used due to its high hardness and strength. The reason behind the high hardness and strength is a large percentage of chromium. This strength and hardness are uniforms throughout the workpiece due to molybdenum. This metal has high tensile strength as compared to other alloy steel in this grade. AISI 4040 (alloy steel) is commonly used as a material for fabricating shaft, gear, bolt, crankshaft, dies, etc.

1.3.4 THE POWDER USED

The response obtained in pwEDM depends on the properties of the powder, that is, the physical, electrical, and thermal properties of the powder. The powder used for this study is graphite powder of size 25 micron, that is, (550-mesh size). When compared with other powder it is clearly seen that graphite powder has low density, high thermal conductivity, and high electrical conductivity as compared to other powders like nickel, titanium, tungsten which is shown in Table 1.2.

TABLE 1.2 Properties of Different Powders

Material	Density (gm/cm^3)	Electrical Resistivity (μ/cm)	Thermal Conductivity (W/mK)
Aluminum	2.70	2.89	236
Chromium	7.16	2.60	95
Copper	8.96	1.71	401
Graphite	1.26	103	3,000
Nickel	8.91	9.5	94
Silicon	2.33	2,325	168
Titanium	4.72	47	22
Tungsten	19.25	5.3	182
Alumina	3.98	103	39.1
CNTs	2.00	50	4,000
Molybdenum disulfide	5.06	106	138
Silicon carbide	3.22	1,013	300

The main reason for selecting graphite powder is density. The main challenge in pwEDM is when the powder is mixed with the electrolyte solution, owing to the density or weight powder goes down and settles at the bottom of the tank, which contains the electrolytic solution. But being low in density the problem of settling down of the powder particles is avoided while using graphite powder. Continuous stirring is needed to get the best result in pwEDM. The graphite powder has high thermal conductivity and low density. Due to the high thermal conductivity of graphite, as compared to other powders, the heat developed by spark energy in the cutting area is dissipated because of which the crater size developed on the surface is smaller which helps in improving the surface finish.

1.3.5 SELECTION OF THE PROCESS PARAMETERS

In this experimentation, input factors considered for powder mixed WEDM are T_{ON}, T_{OFF}, SV, WF, and PC response variables are Ra, and MRR. It is shown in Table 1.3.

The different parameters and their levels are shown in Table 1.4.

TABLE 1.3 Process Input Factors and Response (Output) Variables

Process Input Factors	Response (Output) Variables
Ton (μs)	MRR (g/min)
Toff (μs)	Ra (μm)
SV (volt)	SG (μm)
WF (m/min)	
Powder conc. (g/l)	

TABLE 1.4 Different Parameters and their Levels

Factors Levels	Ton (μs)	Toff (μs)	SV (V)	WF (m/min)	PC (g/l)
1	115	45	10	2	0
2	120	50	20	4	0.2
3	125	55	30	8	0.3

On the basis of the selection of process parameters and their level, the orthogonal array is generated. In this, case, all input parameters are considered at three levels so that a L27 orthogonal array has been selected. The design matrix is shown in Table 1.5.

TABLE 1.5 Design Matrix as per Taguchi L27 Array

Sr. No.	T_{ON}	T_{OFF}	SV	WF	PC
1.	115	45	10	2	0
2.	115	45	10	2	0.2
3.	115	45	10	2	0.3
4.	115	50	20	4	0
5.	115	50	20	4	0.2
6.	115	50	20	4	0.3
7.	115	55	30	8	0
8.	115	55	30	8	0.2
9.	115	55	30	8	0.3
10.	120	45	20	8	0
11.	120	45	20	8	0.2
12.	120	45	20	8	0.3
13.	120	50	30	2	0

TABLE 1.5 *(Continued)*

Sr. No.	T_{ON}	T_{OFF}	SV	WF	PC
14.	120	50	30	2	0.2
15.	120	50	30	2	0.3
16.	120	55	10	4	0
17.	120	55	10	4	0.2
18.	120	55	10	4	0.3
19.	125	45	30	4	0
20.	125	45	30	4	0.2
21.	125	45	30	4	0.3
22.	125	50	10	8	0
23.	125	50	10	8	0.2
24.	125	50	10	8	0.3
25.	125	55	20	2	0
26.	125	55	20	2	0.2
27.	125	55	20	2	0.3

1.3.6 INSTRUMENTATIONS FOR CARRYING OUT THE MEASUREMENTS OF MRR AND SR

1.3.6.1 MATERIAL REMOVAL RATE MEASUREMENT

Material removal rate can be defined as "the amount of material, that is, removed from the workpiece during the cutting process per unit time." In this study, the material removal rate is measured in 'g/min.' To calculate the material removal rate following formulae is used.

$$MRR = \frac{\text{Weight of w/p before machining (gm)- Weight of w/p after machining (gm)}}{\text{Time required for machining (min)}} \quad (1.7)$$

The test specimen is weighted before the cutting and after cutting. The weight of the specimen was carried out by using the POSCO weighing machine having the least count of 0.001 gm. Cutting time is calculated by using a stopwatch.

1.3.6.2 SURFACE ROUGHNESS (RA) MEASUREMENT

The Ra value is measured with 'Mitutoyo surface roughness tester SJ 210' at GCE, Karad. The "p" indicates profile developed on any section of the specimen after wire electric discharge machining. This profile is generally obtained by moving the probe of surface roughness tester perpendicular to the machined surface. The "l" represents sampling length, which is the length of the profile used for the calculation of Ra value. In general, roughness is measured in terms of average roughness (Ra).

1.4 RESULTS AND DISCUSSIONS

Experimentations have been carried out as per the design matrix and the results are recorded, as per the descriptions given in the previous section. The outputs corresponding to the design matrix are recorded in Table 1.6.

TABLE 1.6 Outputs of the Experimentation

Sr. No.	T_{ON}	T_{OFF}	SV	WF	PC	MRR	SR
1.	115	45	10	2	0	0.023140	3.554
2.	115	45	10	2	0.2	0.020737	2.658
3.	115	45	10	2	0.3	0.019955	2.617
4.	115	50	20	4	0	0.018300	2.633
5.	115	50	20	4	0.2	0.013526	2.366
6.	115	50	20	4	0.3	0.011847	2.328
7.	115	55	30	8	0	0.014408	2.497
8.	115	55	30	8	0.2	0.009746	2.335
9.	115	55	30	8	0.3	0.008234	1.874
10.	120	45	20	8	0	0.027096	3.326
11.	120	45	20	8	0.2	0.020095	2.661
12.	120	45	20	8	0.3	0.018947	2.546
13.	120	50	30	2	0	0.021105	3.193
14.	120	50	30	2	0.2	0.013312	2.295
15.	120	50	30	2	0.3	0.013196	2.250
16.	120	55	10	4	0	0.031075	3.800
17.	120	55	10	4	0.2	0.025377	3.097

TABLE 1.6 *(Continued)*

Sr. No.	T_{ON}	T_{OFF}	SV	WF	PC	MRR	SR
18.	120	55	10	4	0.3	0.023661	2.823
19.	125	45	30	4	0	0.029065	3.975
20.	125	45	30	4	0.2	0.021791	2.585
21.	125	45	30	4	0.3	0.021377	2.551
22.	125	50	10	8	0	0.042424	4.056
23.	125	50	10	8	0.2	0.036842	3.788
24.	125	50	10	8	0.3	0.032098	3.567
25.	125	55	20	2	0	0.032558	3.796
26.	125	55	20	2	0.2	0.028390	3.125
27.	125	55	20	2	0.3	0.024861	3.034

1.4.1 *SURFACE ROUGHNESS*

1.4.1.1 *TAGUCHI ANALYSIS FOR SURFACE ROUGHNESS*

The average values of surface roughness at various levels of pulse on time, pulse off time, servo voltage, wire feed rate, and powder concentration is shown in Figure 1.4.

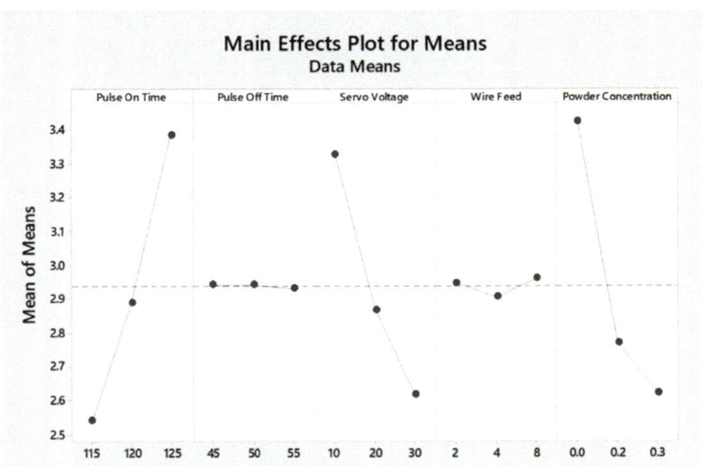

FIGURE 1.4 The main effect plot for Ra data means v/s pulse on time, pulse off time, servo voltage, wire feed, powder concentration.

It indicates that the level 1 of pulse on time, that is, 115 µs gives the lowest average Ra value 2.540 µm whereas the level 2, that is, 120 µs gives an average Ra value 2.888 µm. The level 3, that is, 125 µs gives maximum average Ra value 3.386 µm. Hence, it is clear that the pulse on time should be kept at level 1 in order to have lower average Ra values.

In the case of pulse off time, level 1 of pulse off time, that is, 45 µs gives an average Ra value 2.941 µm whereas the level 2, that is, 50 µs gives maximum average Ra value 2.942 µm and level 3, that is, 55 µs gives minimum average Ra value 2.931 µm, so for minimum Ra pulse off time must be set to level 3.

In case of servo voltage, level 1, that is, 10 V gives maximum average Ra value, that is, 3.329 µm, level 2 of servo voltage, that is, 20 V gives average Ra value 2.868 µm and level 3 of servo voltage, that is, 30 V gives lowest average Ra value, that is, 2.617 µm. Therefore, to have a lower average value of Ra servo voltage should be at level 3.

In the case of wire feed rate, level 1 of wire feed, that is, 2 m/min gives average Ra value 2.947 µm and level 2, that is, 4 m/min gives the lowest Ra value 2.906 µm, level 3, that is, 8 m/min gives the highest value of average Ra 2.961 µm. So that the wire feed rate should keep at level 2 to have a lower average value of Ra.

In the case of powder concentration, level 1 of powder concentration, that is, 0 g/l gives the highest average Ra value = 3.426 µm and level 2, that is, 0.1 g/l gives average Ra value 2.768 µm, level 3, that is, 0.3 g/l gives the lowest value of average Ra 2.621 µm. So that the powder concentration rate should keep at level 3 to have a lower average value of Ra.

The responses obtained for the means of surface roughness obtained by the Taguchi method is given in Table 1.7.

TABLE 1.7 Response Table for Means of the Surface Roughness

Level	Ton	Toff	SV	WF	PC
1	2.540*	2.941	3.329	2.947	3.426
2	2.888	2.942	2.868	**2.906***	2.768
3	3.386	**2.931***	**2.617***	2.961	**2.621***
Delta	0.846	0.011	0.712	0.055	0.804
Rank	1	5	3	4	2

From the main effect plot of surface roughness, it's clear that for achieving the minimum value of surface roughness pulse on time is set at 115 μs. The main reason behind it is that, at the lower value of pulse on time, less amount of spark is generated resulting in less amount of spark energy. Due to this, the crater size on the surface, which is responsible for surface roughness, is decreased as compared to the size of the crater at a higher value of pulse on time. This is the main reason for surface roughness being minimum at a lower value of pulse on time.

At the higher value of pulse off time, less spark energy developed, this helps to improve the surface roughness value. Hence, in that case, surface roughness is minimum at a higher value of pulse off time. Also, as can be seen from Figure 1.4, there is not much variation in surface roughness with pulse off time and it remains more or less constant with only a small decrease in surface roughness at higher values due to reasons as explained.

Servo voltage controls the gap between the electrode and the workpiece. When the servo voltage is high, it basically means that the gap between the workpiece and electrode is more. Due to this, the interaction between the spark and the workpiece is reduced; hence the crater size developed on the surface due to fewer is small, which results in lower surface roughness.

At high cutting speed if the wire feed rate is low the repeated breakage of wire takes place and if the wire feed rate is more than the optimum wire consumption is more which increases the production cost. Because of that, it is necessary to set the wire feed rate at an appropriate value. Due to this reason, the result shows the wire feed rate optimum at 4 m/min.

As graphite powder concentration increases, it carries away more heat from the sparking zone due to the high thermal conductivity of graphite. Due to this at a higher concentration of graphite powder, the surface roughness is reduced (or the surface finish obtained is very good).

1.4.1.2 ANALYSIS OF VARIANCE (ANOVA) ANALYSIS

ANOVA is a statistical technique used for calculating the percentage contribution of each process parameter in the response. After carrying out the ANOVA analysis, it is clearly found which parameters have a higher influence on the particular output (surface roughness) in this case. Once it is established, special care can be taken about that parameter which is having the highest influence. Analysis of variance for Ra has been done by considering Taguchi L27 orthogonal array. ANOVA is carried out in Mini Tab-18, which is used to find the significant and sub significant parameters for Ra.

ANOVA analysis for data means is shown in Table 1.8. It indicates that the powder concentration is the most influential parameter on surface roughness with 34.22% followed by pulse on-time 33.72% and servo voltage with 24.28%. Wire feed rate has a very low contribution of 0.15% and pulses off time have an almost negligible contribution of 0.010%.

TABLE 1.8 Response Table for Means of Surface Roughness by ANOVA

Source	DF	Seq SS	Adj SS	Adj MS	F-Value	P-Value	Contribution
Ton	2	3.2556	3.2556	1.62783	35.38	0.000	33.72%
Toff	2	0.0006	0.0006	0.00032	0.01	0.993	0.01%
SV	2	2.3449	2.3449	1.17246	25.48	0.000	24.28%
WF	2	0.0144	0.0144	0.00724	0.16	0.856	0.15%
PC	2	3.3039	3.3039	1.65197	35.90	0.000	34.22%
Error	16	0.7362	0.7362	0.04601			7.62%
Total	26	9.6558					100.00%

The percentage contribution of each factor is graphically shown in Figure 1.5.

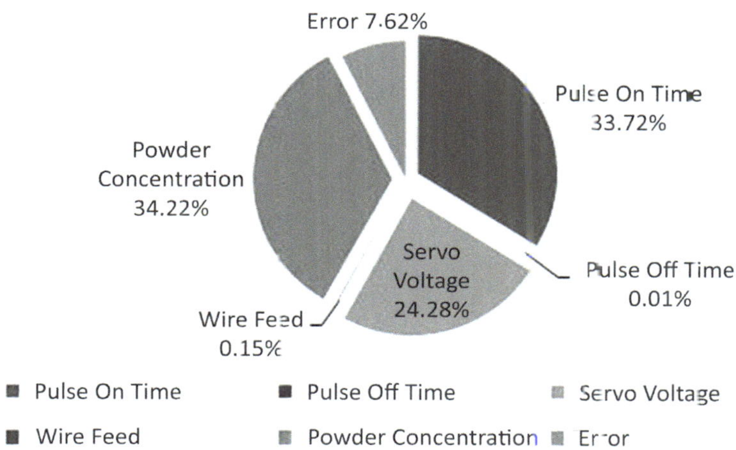

FIGURE 1.5 Percentage contribution of each factor on surface roughness.

1.4.1.3 S/N RATIO FOR SURFACE ROUGHNESS

When large data is available for analysis, it is preferable to use MSD (mean square deviation). It combines the results of average deviation and SD (standard deviation). For analysis of large data, logarithmic transformation of mean square deviation is taken, which is called the signal to noise ratio. The signal is a desirable quantity and noise is an undesirable quantity so its ratio is considered at the maximum value it means that a high S/N ratio means it contains low noise value, which is desirable.

To obtain optimal cutting performance, the lower the better quality characteristic for surface roughness must be taken (Figure 1.6).

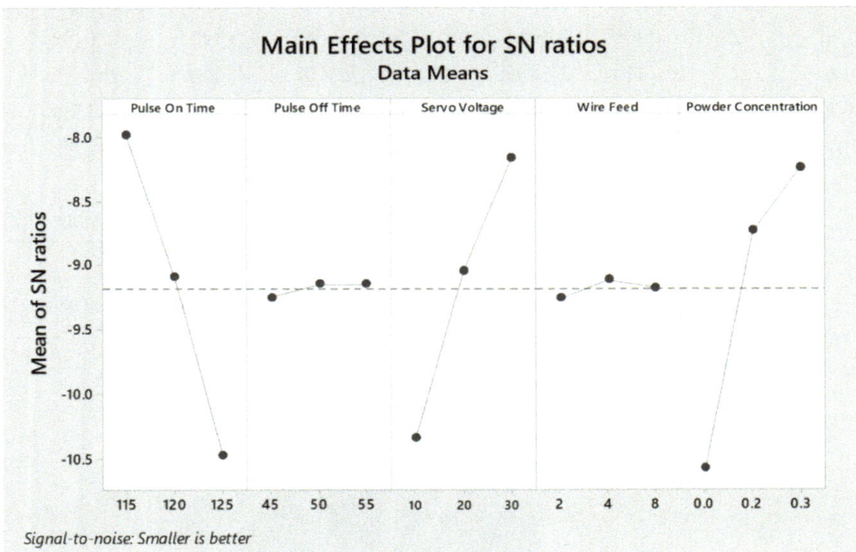

FIGURE 1.6 Main plots for S/N ratio of Ra (L27 array).

The graphs of the main effects are shown in Figure 1.6 the optimum levels for pwEDM parameters can be predicted from the main plots for Ra data S/N ratio. It shows the level number 1 for a pulse on time, level 3 for pulse off time, level 3 for servo voltage, level 2 for wire feed rate, and level 3 for powder concentration, gives the maximum value of S/N ratio.

1.4.2 MATERIAL REMOVAL RATE

1.4.2.1 TAGUCHI ANALYSIS FOR MATERIAL REMOVAL RATE

Figure 1.7 shows the average MRR values at various levels of pulse on time, pulse off time, servo voltage, wire feed rate, and powder concentration. It indicates that the level 1 of pulse on time, that is, 115 μs gives the lowest average MRR value = 0.01554 g/min whereas level 2, that is, 120 μs gives an average MRR value = 0.02154 g/min. Level 3, that is, 125 μs gives the maximum MRR value = 0.02993 g/min, Hence it is clear that pulse on time should be kept at level 3 in order to have higher average MRR values.

In case of pulse off time, the level 1, that is, 45 μs gives the average MRR value = 0.02247 g/min, level 2, that is, 50 μs gives maximum MRR value = 0.02252 g/min and level 3, that is, 55 μs gives lowest average MRR value = 0.02203 g/min. Hence, it is clear that the pulse off-time should be kept at level 2 in order to have higher MRR values. The main effect plot for MRR with the input parameters is shown in Figure 1.7.

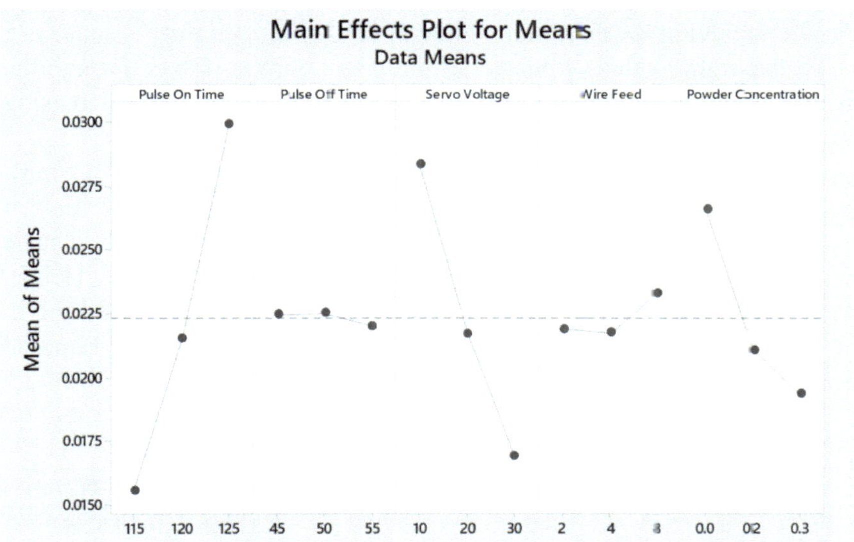

FIGURE 1.7 Main effect plot for MRR data means v/s pulse on time, pulse off time, servo voltage, wire feed rate, powder concentration.

Similarly, the average MRR values at different levels of servo voltage and wire feed rate are shown graphically in Figure 1.7. It is seen that level

1 of the servo voltage gives the maximum average MRR value = 0.02837 g/min, level 2 of servo voltage gives the average MRR value = 0.02174 g/min, and level 3 of servo voltage gives the lower average value of MRR = 0.01691 g/min. Therefore, servo voltage should be kept at its level 1 to have a higher average MRR.

For level 1 of wire feed rate, that is, 2 m/min, there is the average value of MRR = 0.02192 g/min, level 2 of wire feed rate, that is, 4 m/min gives the minimum average value of MRR value = 0.02178 g/min, level 3 of the wire feed rate, that is, 8 m/min gives highest average MRR value = 0.02332 g/min, so that to have the highest MRR wire feed rate should be kept at level 3, that is, 8 m/min.

For level 1 of powder concentration, that is, 0.0 g/l there is a maximum value of MRR = 0.02657 g/min, level 2 of powder concentration, that is, 0.2 g/l gives an average value of MRR value = 0.02109 g/min, level 3 of powder concentration, that is, 0.3 g/l gives lower average MRR value = 0.01935 g/min, so that to have highest MRR powder concentration should be kept at level 1, that is, 0.0 g/l.

From the main effect plot of the material removal rate, it is observed that for achieving the maximum value of MRR pulse on time is set at 125 μs. At the higher value of pulse, on-time more current passes through the electrode and more spark energy are generated which helps in increasing the material removal rate.

As per the theory, the pulse off-time should be minimum and simultaneously the pulse on time should be maximum to ensure the highest material removal rate. But, experimentally it was observed that if the pulse off-time is kept at level 3 frequent breakage of wire takes place. Hence, experimentally it is observed that pulse off time at level 2 is resulting in the highest MRR. The response table for the means of material removal rate is given in Table 1.9.

TABLE 1.9 Response Table for Means of the Material Removal Rate

Level	Ton	Toff	SV	WF	PC
1	0.01554	0.02247	**0.02837***	0.02192	**0.02657***
2	0.02154	**0.02252***	0.02174	0.02178	0.02109
3	**0.02993***	0.02203	0.01691	**0.02332***	0.01935
Delta	0.01439	0.00048	0.01145	0.00154	0.00722
Rank	1	5	2	4	3

At a lower value of the servo, the voltage gap between the electrode and electrode is less, so more spark comes in contact with the workpiece it helps for increasing the material removal rate.

The wire feed rate is to be set at optimum value, that is, 4 m/min. At this level due to high spark energy, if we set the wire feed rate at optimum value, that is, 4 m/min. it was experimentally observed that very frequent breakage of wire takes place during machining, so it's necessary to set the wire feed rate at a higher value to avoid damage to the wire.

For achieving maximum MRR, powder concentration is required to be set at a low-level otherwise more heat will be dissipated by the powder from the cutting zone, which will ultimately decrease the MRR. So experimental result shows that for getting maximum MRR, it's necessary to set at a low level, that is, 0.0 g/l.

1.4.2.2 ANALYSIS OF VARIANCE (ANOVA) ANALYSIS

ANOVA analysis for data means is shown in Table 1.10. It indicates that the pulse on time is the most affecting parameter on material removal rate with 51.41% followed by servo voltage 32.54% and powder concentration with 13.98%. Wire feed rate and pulse off time have a contribution of 0.72% and 0.07%, respectively. ANOVA table with percentage contribution of each factor to the material removal rate gives in Table 1.10.

The percentage contribution of each factor in from a pie graph is shown in Figure 1.8.

TABLE 1.10 ANOVA for Means of Material Removal Rate (L27)

Source	DF	Seq SS	Adj SS	Adj MS	F-Value	P-Value	Contribution
Ton	2	0.000940	0.000940	0.000470	319.38	0.000	51.41%
Toff	2	0.000001	0.000001	0.000001	0.43	0.658	0.07%
SV	2	0.000595	0.000595	0.000298	202.11	0.000	32.54%
WF	2	0.000013	0.000013	0.000007	4.45	0.029	0.72%
PC	2	0.000256	0.000256	0.000128	86.84	0.000	13.98%
Error	16	0.000024	0.000024	0.000001			1.29%
Total	26	0.001829					100.00%

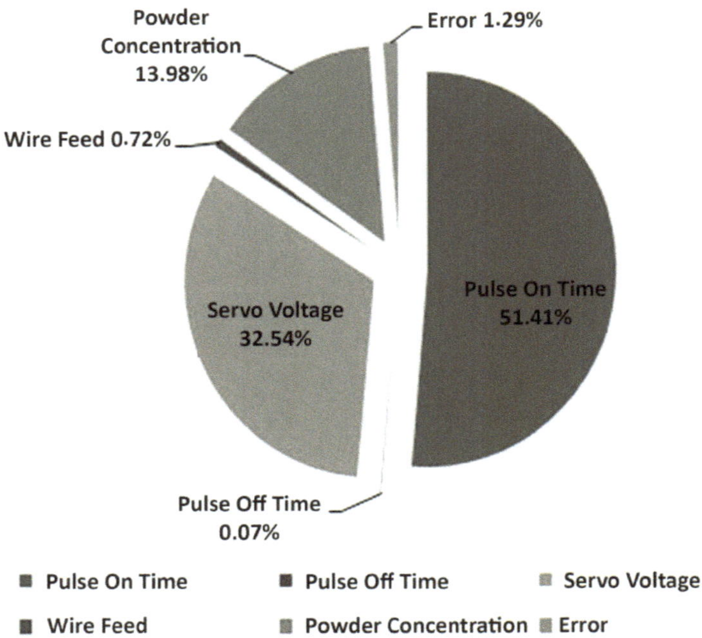

FIGURE 1.8 Percentage contribution of each factor on the material removal rate.

1.4.2.3 S/N RATIO OF MATERIAL REMOVAL RATE

The graphs of the main effects plot are shown in Figure 1.9. The optimum levels for pwWEDM parameters can be predicted from the main plots of S/N ratio for MRR data. It shows the level number 3 for a pulse on time, level 1 for pulse off time, level 1 of servo voltage, level 1 of wire feed rate, and level 1 of powder concentration, gives the maximum value of S/N ratio. For Higher MRR S/N ratio should be higher as shown in the main effect plot for S/N (MRR).

Some parameters are prescribed by the manufacturer of the machine, which are the default parameters used for carrying out the pwEDM process. The outputs obtained from using those parameters are given in Table 1.11. But, the parameters obtained through the Taguchi method yielded much better results as given in Table 1.12.

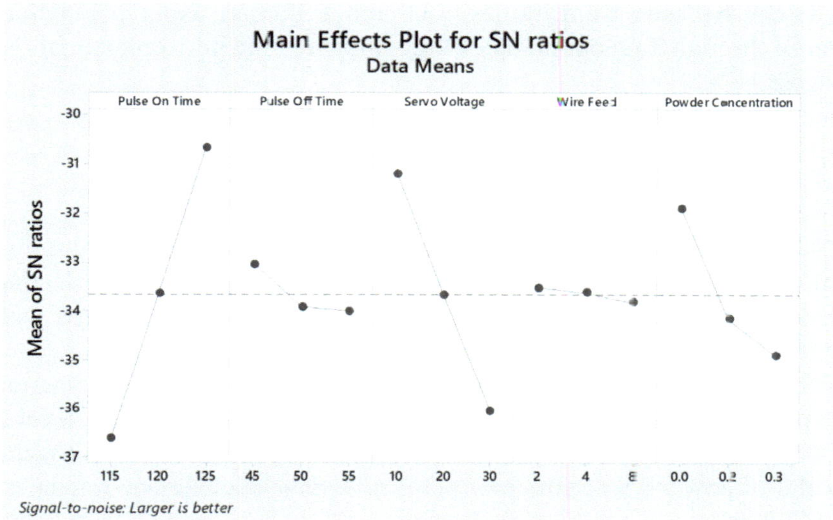

FIGURE 1.9 Main plots for S/N ratio of MRR (L27 array).

TABLE 1.11 Process Parameters at the Initial Level and Values of the Response Variable

Sr. No.	Process Parameters	Ra(µm)		MRR (g/min)	
		Initial Level	**Value**	**Initial Level**	**Value**
1.	T_{on}	1	115	1	115
2.	T_{off}	1	45	1	45
3.	SV	1	10	1	10
4.	WF	1	2	1	2
5.	PC	1	0.0	1	0.0
	Final Value	3.554 (µm)		0.023140 (g/min)	

TABLE 1.12 Conclusion Based on Taguchi, ANOVA, and Analysis by S/N Ratio

Sr. No.	Process Parameters	Ra(µm)		MRR (g/min)	
		Initial Level	**Value**	**Initial Level**	**Value**
1.	T_{on}	1	115	3	125
2.	T_{off}	3	55	1	45
3.	SV	3	30	1	10
4.	WF	2	4	1	2
5.	PC	3	0.3	1	0
	Final Value	1.792 (µm)		0.048647 (g/min)	

It can be seen from Tables 1.11 and 1.12 that using parameters chosen through Taguchi methodology improves the performance of the process.

1.5 CONCLUSIONS AND FUTURE SCOPE

ANOVA for data means of Ra obtained after pwEDM of AISI 4140 shows that the powder concentration is the most significant parameter affecting the Ra with 34.22% followed by T_{ON} 33.72% and servo voltage with 24.28%.

S/N ratio analysis for Ra predicts the optimum parameter combination for pmWEDM parameters as level number 1 for Ton (115 μs), level 3 of Toff (55 μs), level 3 of SV (30 V), level 2 of WF rate (4 m/min), and level 3 of powder concentration (0.3 g/l). At this level surface roughness is 1.792 μm.

ANOVA for data means of MRR getting after pwEDM of AISI 4140 shows that the T_{ON} is the most affecting parameter on MRR with 51.41% followed by servo voltage 32.54% and powder concentration (g/l) with 13.98%.

S/N ratio analysis for MRR predicts an optimum parameter combination for pwEDM as level number 3 of Ton (125 μs), level 1 of Toff (45 μs), level 1 of SV (10 V), level 1 of WF rate (2 m/min) and level 1 of powder concentration (1 g/l). At this level, MRR is 0.048647 (g/min).

When the input process parameters are chosen through ANOVA and S/N analysis the surface roughness gets reduced from 3.554 to 1.792, which is a reduction of 49.58%. MRR is increased from 0.023 g/min to 0.048 g/min, which is an increase of 73.91%. So, it can be concluded that through the application of ANOVA and S/N analysis, process parameters can be chosen to improve the desired output.

ACKNOWLEDGMENT

The authors would like to acknowledge Government Engineering College, Karad for providing the research facilities.

KEYWORDS

- **AISI 4140**
- **analysis of variance**
- **L27 orthogonal array**
- **material removal rate**
- **powder-assisted wire electric discharge machining**
- **surface roughness**
- **Taguchi method**

REFERENCES

1. Roche, H., Delagnes, A., Brugal, J. P., Feibel, C., Kiburjia, M., Mourre, V., & Texier, P. J., (1999). Early hominid stone tool production and technical skill 2.34 M year ago in West Turkana, Kenya. *Nature, 399*(6731), 57–60.
2. Dauw, D., (1989). *Charmilles Technologies: Facing the Future*. In a paper presented during the inauguration of CT-Japan.
3. Stovicek, D. R., (1993). The state-of-the-art EDM science. *Journal of Tooling and Production, 59*(2), 1–42.
4. Singh, S., Maheshwari, S., & Pandey, P. C., (2004). Some investigations into the electric discharge machining of hardened tool steel using different electrode materials. *Journal of Materials Processing Technology, 149*, 272–277.
5. Chen, S. L., Yan, B. H., & Huang, F. Y., (1999). Influence of kerosene and distilled water as dielectrics on the electric discharge machining characteristics of Ti-6A1-4V. *Journal of Materials Processing Technology, 87*, 107–111.
6. Boujelbene, M., Bayraktar, E., Tebni, W., & Salem, S. B., (2009). Influence of machining parameters on the surface integrity in electrical discharge machining. *Archives of Materials Science Engineering, 37*(2), 110–116.
7. Ugrasen, G., Singh, M. R. B., & Ravindra, H. V., (2018). Optimization of process parameters for SS304 in wire electrical discharge machining using Taguchi's technique. *Materials Today: Proceedings, 5*(1), 2877–2883.
8. Goswami, A., & Kumar, J., (2017). Trim cut machining and surface integrity analysis of nimonic 80 A alloy using wire cut EDM. *Engineering Science Technology: An International Journal, 20*(1), 175–186.
9. Kumar, V., Sharma, N., Kumar, K., & Khanna, R., (2018). Surface modification of WC-Co alloy using Al and Si powder through WEDM: A thermal erosion process. *Particulate Science and Technology, 36*(7), 878–886.
10. Kansal, H. K., Singh, S., & Kumar, P., (2007). Technology and research developments in powder mixed electric discharge machining (PMEDM). *Journal of Materials Processing Technology, 184*, 32–41.

11. Batish, A., Bhattacharya, A., Singla, V. K., & Singh, G., (2012). Study of material transfer mechanism in die steels using powder mixed electric discharge machining. *Materials and Manufacturing Processes*, *27*(4), 449–456.
12. Prakash, C., Kansal, H. K., Pablas, B. S., & Puri, S., (2017). Experimental investigations in powder mixed electric discharge machining of Ti-35Nb-7Ta-5Zrβ-titanium alloy. *Materials and Manufacturing Processes, 32*(3), 274–285.
13. Talla, G., Sahoo, D. K., Gangopadhyay, S., & Biswas, C. K., (2015). Modeling and multi-objective optimization of powder mixed electric discharge machining process of aluminum/alumina metal matrix composite. *Engineering Science Technology an International Journal, 18*(3), 369–373.
14. Kumar, A., Maheshwari, S., Sharma, C., & Beri, N., (2011). Analysis of machining characteristics in additive mixed electric discharge machining of nickel-based superalloy Inconel 718. *Materials and Manufacturing Processes, 26*(8), 1011–1018.
15. Kansal, H. K., Singh, S., & Kumar, P., (2006). Performance parameters optimization (multi-characteristics) of powder mixed electric discharge machining (PMEDM) through Taguchi's method and utility concept. *Indian Journal of Engineering Sciences and Material Science, 13*, 209–216.
16. Kansal, H. K., Singh, S., & Kumar, P., (2005). Parametric optimization of powder mixed electrical discharge machining by response surface methodology. *Journal of Materials Processing Technology, 169*(3), 427–436.
17. Kumar, V., Sharma, N., Kumar, K., & Khanna, R., (2018). Surface modification of WC-Co alloy using Al and Si powder through WEDM: A thermal erosion process. *Particulate Science and Technology, 36*(7), 878–886.
18. Jarin, S., Saleh, T., Rana, M., Muthalif, A. G. A., & Ali, M. Y., (2017). An experimental investigation on the effect of nanopowder for micro-wire electro-discharge machining of gold-coated silicon. *Procedia Engineering, 184*, 171–177.

CHAPTER 2

Three and Four Precision Position Graphical and Analytical Synthesis Procedure Mechanism Design for Agricultural Tillage Operation

N. R. N. V. GOWRIPATHI RAO,[1] HIMANSHU CHAUDHARY,[1] and AJAY KUMAR SHARMA[2]

[1]*Malaviya National Institute of Technology Jaipur, Rajasthan – 302017, India, E-mail: gowripathiraofmpe@gmail.com*

[2]*College of Technology and Engineering, MPUAT Udaipur, Rajasthan – 313001, India*

ABSTRACT

Agricultural tillage plays an essential role in the farming community. There are different agricultural operations which contribute to the overall development of the crop, but among them, agricultural tillage plays an important role. Various unit operations are tillage, sowing and fertilizing, irrigation, harvesting, and post-processing activities. The chapter deals to design a four-bar mechanism for a concept known as vibratory tillage, which can be used for agricultural soil manipulation operation with improved soil properties and less power consumption. Mechanism design is through path generation process and graphical, analytical procedures are used to design a mechanism for a particular vibratory tillage tool trajectory in soil. Three and four precision methods are used to design a four-bar mechanism through which the tool follows the path very accurately. The mechanism design is validated through MATLAB, and it is confirmed that the tool passes through the selected precision points correctly. Also, it is found that the mechanism dimensions obtained through the graphical and analytical techniques are exact.

2.1 INTRODUCTION

India is an agricultural country, and the maximum population is the country that relies on agriculture as its primary source of income. Around 54.6% of the population is engaged in agriculture and other related activities. Indian farmers have significant landholding capacity in the small and marginal segment, which is 1–2 hectares, and around 85% lie in this segment [1, 2]. Nowadays there is an increased demand for food due to population rise. Thus, there is a need to improve the farm mechanization, which can help small and marginal farmers cultivate better and contribute to nation-building and farm productivity.

The present level of farm mechanization status in India is (2.24 kW/ha) (2016–17) [3], and most of the agricultural operations in India are performed through conventional methods. The power availability through the tractor is underutilized by the Indian farmers, which results in the reduced efficiency of the system with increased fuel consumption. There are many issues and problems with the usage of conventional tillage practices which is uneconomical to the farmers. Thus looking into the problem of less farm holding of Indian farmers and utilization of traditional tillage practices there is a need to design and develop an active tillage implement which can perform better and increase the overall economy of the farmers with better utilization of power available from the tractor and less fuel consumption.

Tillage is the mechanical manipulation of the soil and plays a vital role in Indian agricultural operations. Timeliness of operations by Indian farmers has gained importance due to improved agricultural production and productivity. There are different operations such as plowing, harrowing, transplanting, weeding, irrigation, harvesting, and threshing, but among them, tillage is an essential unit operation that needs to take care of crop development. There is a 40% scope of improvement in tillage practices according to Ref. [1].

Tillage is classified into two types one is primary, and another is secondary tillage operation. Primary operation results in the proper physical condition of the soil and secondary operation for better soil tilth. The initial depth of the soil in primary tillage is 20–25 cm and for secondary tillage operation is 10–15 cm, respectively. Mold board plow, disc plow, and subsoiler are the example of primary tillage implements and cultivator, disc harrow is secondary tillage equipment [27]. Nowadays, there is a shift of interest among the farming community to work on the low draft and more efficient power usage agricultural implements. Thus, there is a need to work and put a continuous effort to work on the low draft and power equipment and develop a soil engaging tool that can contribute to completing the work in a single run by the tractor in the fields. Thus, the low draft tillage tool consists of active tillage equipment or combination tillage tool [28, 29]. The main concept of active tillage tool is power supplied through the power take-off shaft (PTO) of the tractor to the tillage cutting tool. Active devices play an important role to achieve savings by coordinating with improved soil properties and energy reduction [20, 21].

Figure 2.1 shows the four-bar mechanism in which the cutting edge of the tool is attached to the coupler end for the soil manipulation process. Thus, inactive tillage tools vibratory tillage plays an important role.

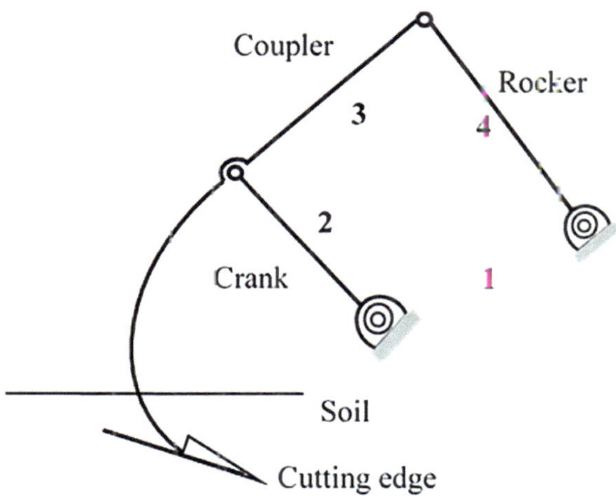

FIGURE 2.1 Active tillage tool [4].

2.1.1 VIBRATORY TILLAGE CONCEPT

The vibratory tillage concept came long back during the 1950s by Ref. [5] and were the first researchers who got the expertise in the area of applying the vibration to the soil working tools. Thus, this encouraged the researchers to work on active tillage systems [22]. Vibrations may be harmonic, simple, and compound depending upon the kind of the mechanism attached. Vibrating tools may have a linear or arc motion with an implement reference frame. Tools oscillate in a particular mode of oscillation with an implement reference frame. This is called a vibratory tillage concept. Inactive vibratory tillage machines motion is given to the tool through the tractor PTO shaft and equipped with a crank mechanism. For speed, reduction purposes the gearbox is provided to reduce the PTO rpm. There are several advantages of vibrating mode tools over a non-oscillating one.

Vibrating tillage tools have several advantages such as reduced draft requirements, improved soil properties [6–8, 23], soil compaction is also reduced due to decreased traction requirements [6, 9–12]. The draft requirement is reduced around 50–60% as compared to non-oscillating one [5, 13, 14, 21]. There is a conflict regarding the total power requirement of the tillage tool. It may increase, decrease, or remain the same for the tillage operation. Some researchers reported that there is an increase of 30–35% power consumption while using the oscillatory tool [15, 16].

Generally, to provide oscillation to the tillage tool there is a need for four-bar mechanisms as shown in Figure 2.2 due to its structural stability and many other advantages. To identify the dimensions kinematic synthesis plays an important role.

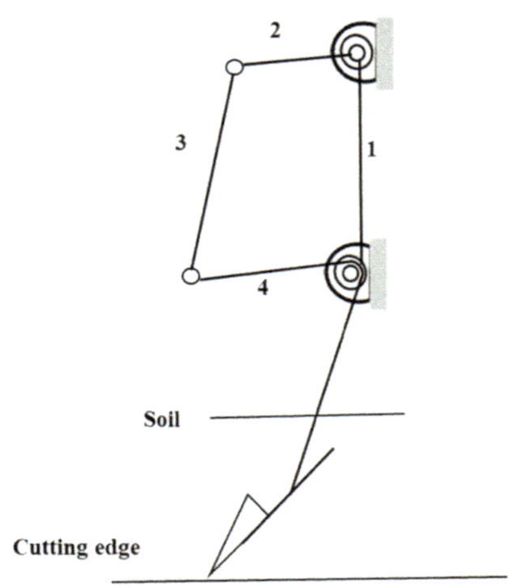

FIGURE 2.2 A four-bar vibratory mechanism in tillage tool [7].

2.2 KINEMATIC SYNTHESIS

Kinematic synthesis plays a vital role to calculate the mechanical dimensions for the output parameters [23, 25, 26]. The basic definition of kinematic synthesis is to design a mechanism for a prescribed output motion for the input motion as shown in Figure 2.4. Mechanism dimensions include link lengths, angles, etc., to achieve the precision position. This is called dimensional synthesis. The reverse process is called analysis as shown in Figure 2.3. There are different types of categories in kinematic syntheses, such as path, motion, and function generation. Synthesis includes graphical and analytical methods. Graphical methods do not include computational effort to solve the problem, and analytical methods include the mathematical models which can be solved analytically and numerically with different algorithms.

The type of problems is defined as a path, motion, and function generation as shown in Figure 2.5. The rigid body in the mechanism designed passes through all the prescribed precision points in the path generation problem and in the motion generation problem mechanism rigid body is guided through prescribed precision positions which are also called rigid body guidance. Function generation deals with the mechanism dimensions to reach the specified link angles.

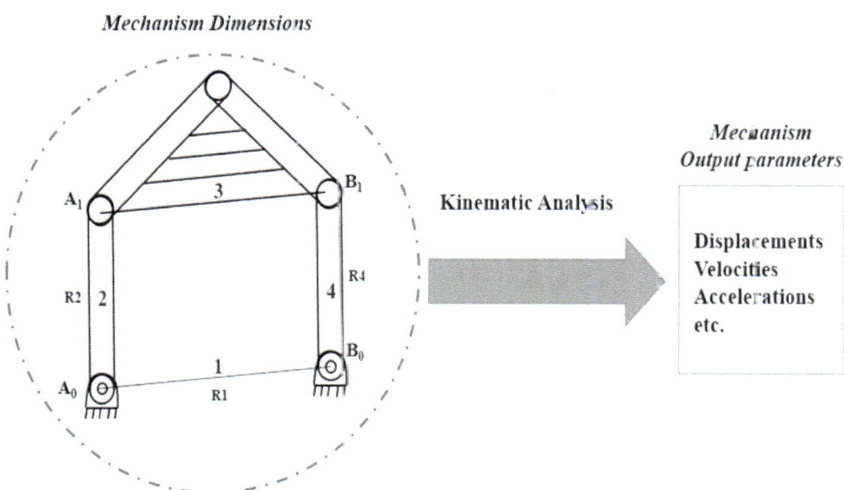

FIGURE 2.3 Kinematic analysis [17].

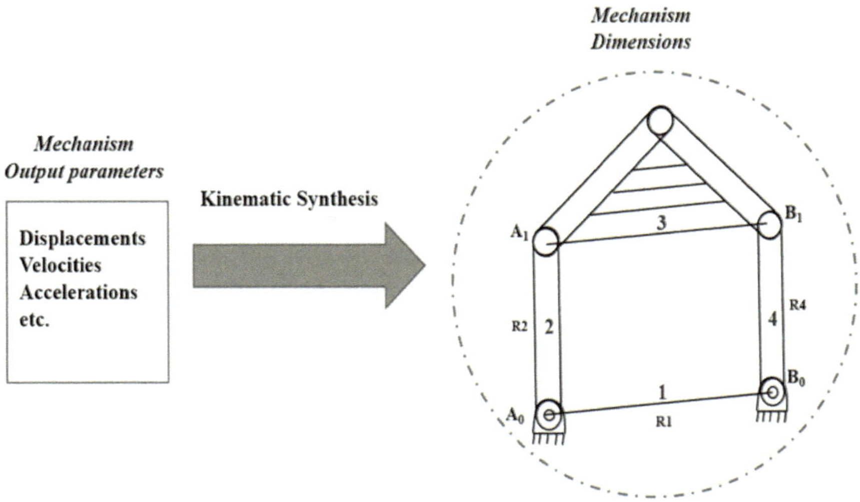

FIGURE 2.4 Kinematic synthesis [17].

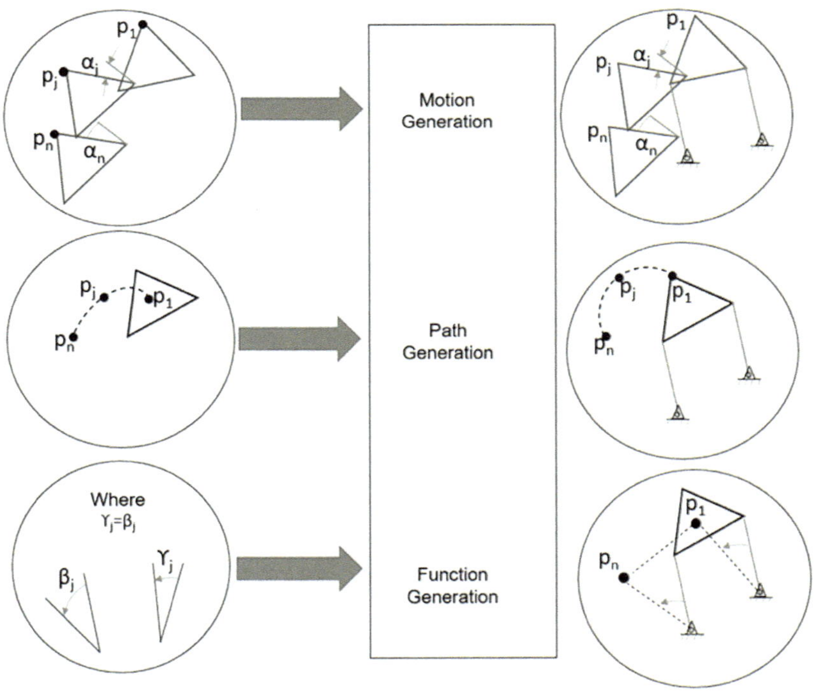

FIGURE 2.5 Different types of mechanism synthesis problems [17].

2.2.1 TOOL TRAJECTORY FOR VIBRATORY TILLAGE

In vibratory tillage operation, the tool has to follow the different phases in the soil to complete the operation. Figure 2.6 shows the tool trajectory in the soil. There are different phases in the soil according to the Ref. [18] for vibratory tillage operation that the tool needs to follow:

1. Cutting of the soil;
2. Backing off phase;
3. Catching up phase;
4. End of the cycle.

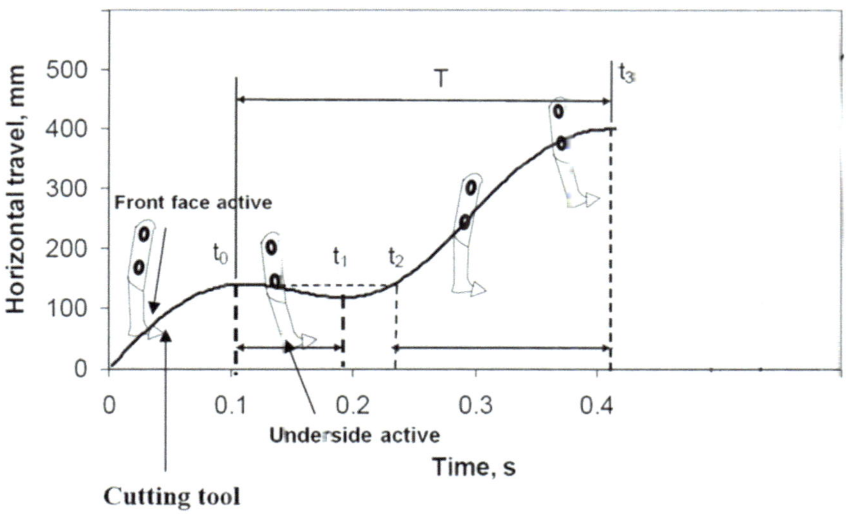

FIGURE 2.6 Tool trajectory in the vibratory tillage operation [18].

- **Cutting Phase:** The part of the oscillation cycle during which the tine tip and the front face penetrates into the uncut soil is called a cutting phase. The tool is oriented at a rake angle of 36^0 normally.
- **Backing Off Phase:** The part of the oscillation cycle at which the tine's front face disengages from uncut soil and the tine goes backward through the loosened soil.
- **Catching Up:** The part of the oscillation cycle for negative and the zero oscillation angles where the tool is moving forward through

the loosened soil, and it continues until it reaches the uncut soil part at the start of the cutting phase. This phase does not exist for positive oscillation angles.

- **End of the Cycle:** This is the ending phase when the cutting tool reaches the loosen soil. Thus, the tool orientation in the soil is completed by following this process. Thus, the designed mechanism should follow the different stages during operation in the soil.

The soil cutting phases by the tillage tools are explained similarly in Refs. [24, 25].

Through the plot digitizer, the coordinates of the trajectory were extracted and for the trajectory available in Figure 2.6. The values of X and Y obtained for the vibratory tillage trajectory are shown in Table 2.1.

TABLE 2.1 Desired Precision Points for the Tool Trajectory [18]

Desired Points	Pxd	Pyd
1	16.741	4.089
2	30.134	7.101
3	40.179	11.214
4	54.688	14.292
5	68.080	19.363
6	84.821	23.452
7	104.911	27.560
8	126.116	31.673
9	141.741	34.756
10	165.179	38.881
11	180.804	33.964
12	194.196	28.036
13	202.009	16.077

The selected values of the tool trajectory for the different phases are explained initially. They are highlighted through bubbles as shown in Figure 2.7. The tool trajectory for the cutting phase is only considered during the study.

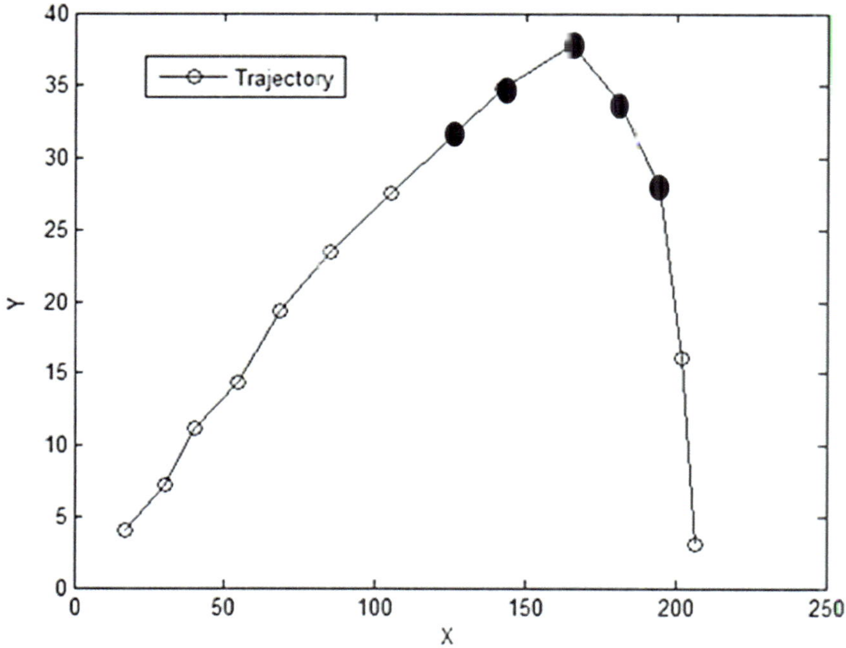

FIGURE 2.7 Tool trajectory in cutting mode.

2.2.2 GRAPHICAL SYNTHESIS OF LINKAGES

A simple geometric procedure proposed in Ref. [19] is adopted there to design a four-bar path generator mechanism which can trace the three precision positions precisely. The objective is to trace the selected three precision points P1, P2, P3 through the coupler point P.

2.2.2.1 THREE PRECISION POSITION GRAPHICAL SYNTHESIS

Three arbitrary precision points of cutting phase are chosen from the experimental trajectory, and the coordinates selected are for the three precision points shown in Table 2.2.

TABLE 2.2 Desired Points of the Tool Trajectory for Three Precision Position [18]

Desired Points (mm)	1	2	3
Pxd	141.741	165.179	180.804
Pyd	34.756	38.881	33.964

In the graphical method, it can be seen that the coupler point P will pass through all precision points as shown in Figure 2.8, but this method does not guarantee a crank-rocker mechanism. The guess method is used to obtain a crank-rocker mechanism. The link lengths obtained from the graphical way are given in the table below, and these link lengths calculated are from the mechanism synthesized through the graphical method. Using the theory of three precision position graphical synthesis methods, the following dimensions were obtained for the four-bar mechanism as shown in Figure 2.9.

AD = r_1 = 370.324 mm;
AB = r_2 = 210.552 mm;
BC = r_3 = 550.237 mm;
CD = r_4 = 340.937 mm;
BP = Coupler length = 280 mm;
CP = Coupler length = 550 mm.

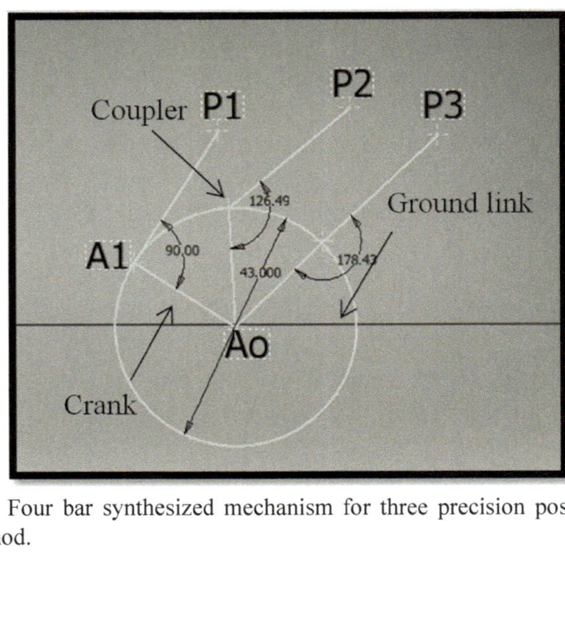

FIGURE 2.8 Four bar synthesized mechanism for three precision positions using the graphical method.

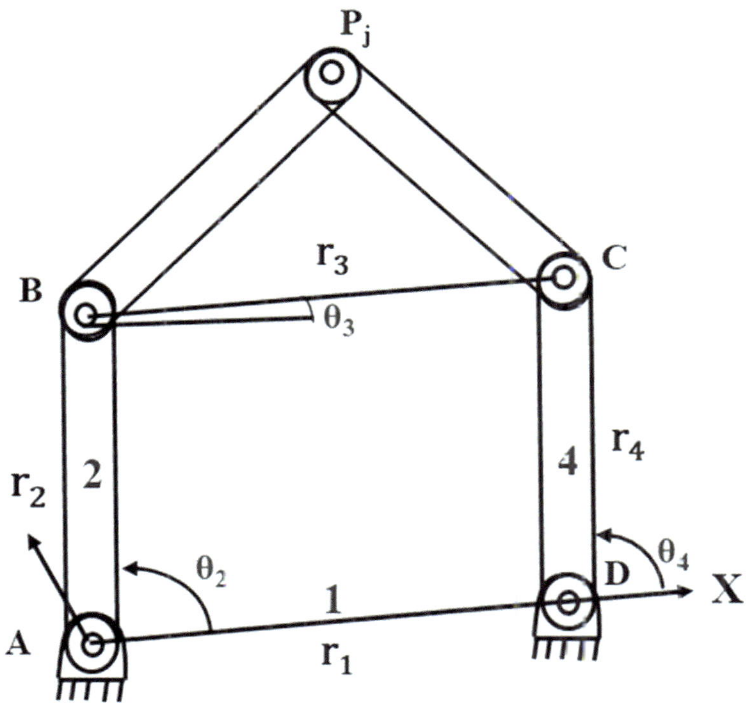

FIGURE 2.9 Synthesized four-bar mechanism with dimensions.

2.2.2.2 FOUR PRECISION POSITION GRAPHICAL SYNTHESIS

A design procedure similar to three precision position synthesis is employed to four precision position using the point position reduction method proposed in Ref. [19] and as shown in Figure 2.10. Without the prescribed timing, the method is applied to four arbitrary points and are chosen from the trajectory studied. The coordinates of the chosen points are given in Table 2.3.

TABLE 2.3 Desired Points of the Tool Trajectory for Four Precision Positions [18]

Desired points (mm)	**1**	**2**	**3**	**4**
Pxd	126.116	141.741	180.804	194.196
Pyd	31.673	34.756	33.954	28.036

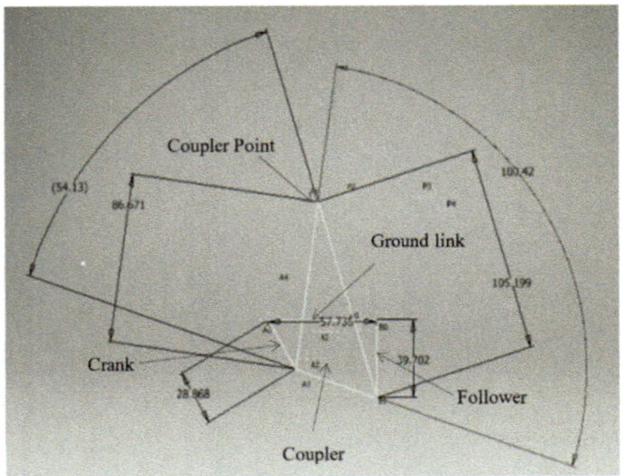

FIGURE 2.10 Four-bar synthesized mechanism for four precision positions using the graphical method.

The link lengths obtained from the graphical method are given below and these link lengths are calculated from the graphical method itself and for the mechanism that is synthesized above. Using the theory of four precision positions graphical synthesis methods the following dimensions were obtained for the four-bar mechanism as shown in Figure 2.11.

AD = r_1 = 570.7350 mm;
AB = r_2 = 280.8675 mm;
BC = r_3 = 450.980 mm;
CD = r_4 = 390.702 mm;
BP = Coupler length = 860.6 mm;
CP = Coupler length = 1050 mm.

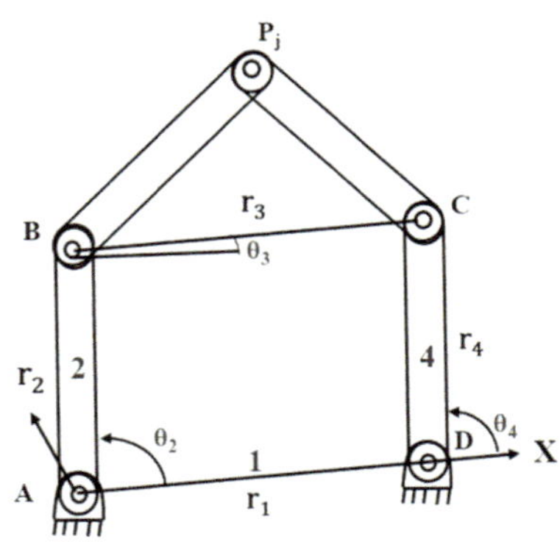

FIGURE 2.11 Synthesized four-bar mechanism with dimensions.

2.2.3 ANALYTICAL LINKAGE SYNTHESIS OF MECHANISM

Linkage synthesis is carried out by three methods defined as geometrical or graphical construction, analytical, and optimal synthesis method. Various mathematical processes to model the planar linkages such as matrix, algebra, and complex number method. But normally complex number technique is considered as the most viable method to solve for linkage synthesis [19]. Three and four precision position mathematical modeling is explained in this section and two examples are considered to explain the process. This section synthesizes a mechanism using the principle theory explained above for three and four precision points. For our tool trajectory, two examples are discussed below for three and four precision points for a path generation problem.

2.2.3.1 THREE PRECISION POSITION ANALYTICAL SYNTHESIS

Path generation synthesis requires precision points for synthesis. Three precision points for the cutting phase were selected for the synthesis procedure from the trajectory [18] and are shown in Figure 2.12. The points are selected as given in Table 2.4.

TABLE 2.4 Desired Four Precision Points for Tool Trajectory [18]

Desired Points (mm)	1	2	3
Pxd	141.741	165.179	180.804
Pyd	34.756	38.881	33.964

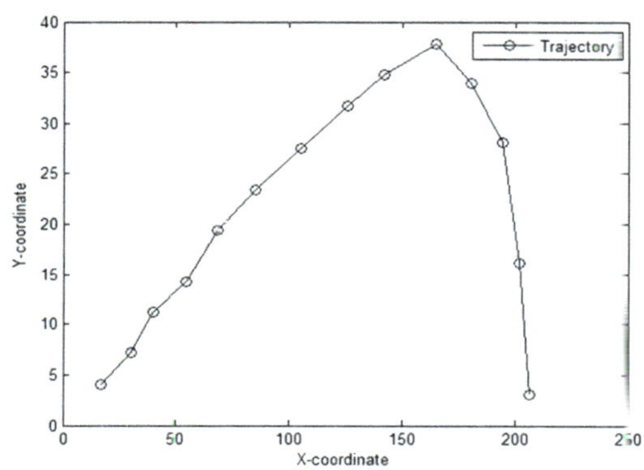

FIGURE 2.12 Actual trajectory and prescribed precision points.

The position vector for the given prescribed coordinates for positions 1, 2, 3 are written as $P_{\delta j} = R_1 - R_j$.

The prescribed values for the following are:

$R_1 = 141.741 + 34.756i$; $R_2 = 168.179 + 38.881i$; $R_3 = 180.804 + 33.964i$

$P_{21} = -26.438 - 4.125i$; $P_{31} = -39.063 + 0.792i$

Using the theory of three precision position analytical synthesis methods (Figure 2.13) the following dimensions were obtained for the four-bar mechanism as shown in Figure 2.14.

$AD = r_1 = 370.324$ mm;

$AB = r_2 = 210.552$ mm;

$BC = r_3 = 550.237$ mm;

$CD = r_4 = 340.937$ mm;

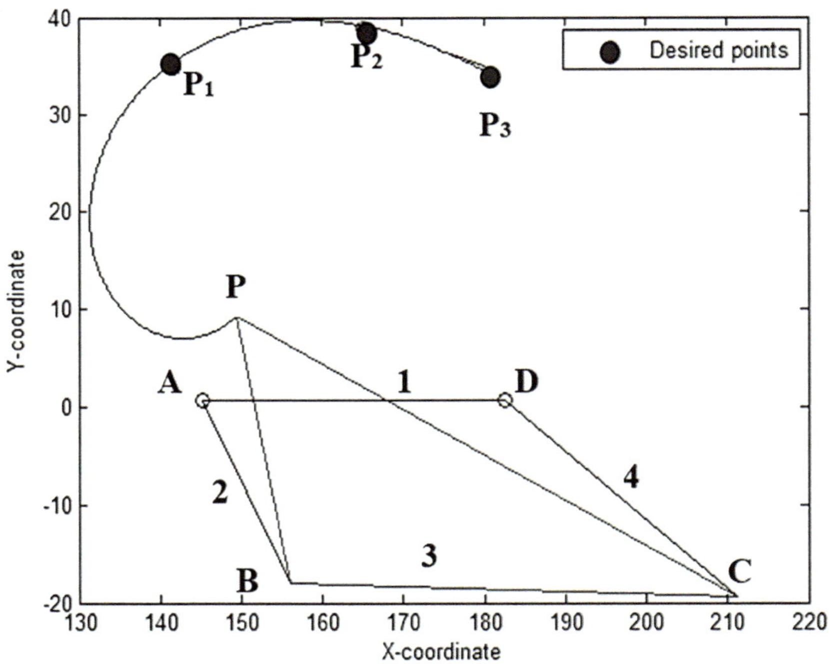

FIGURE 2.13 Synthesized four-bar linkage mechanism for three precision points using the analytical method.

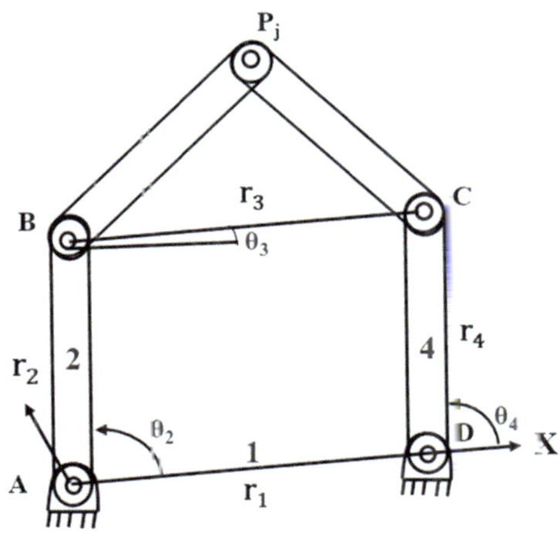

FIGURE 2.14 Synthesized Four bar mechanism with dimensions.

Through the analytical synthesis procedure mechanism, dimensions are obtained and simulated in MATLAB to observe whether the designed mechanism passes through the desired trajectory or not. Figure 2.12 is the actual trajectory through which the mechanism designed should pass through selected precision points. Three precision points were selected from the trajectory. The precision points selected are shown through red-colored circles in Figure 2.12. It is observed that the synthesized four-bar mechanism can easily track the desired three prescribed position points accurately as shown in Figure 2.13. For further accuracy of the path, the tracking number of desired points is increased to four. Synthesis and analysis of four precision points were performed.

2.2.3.2 *FOUR PRECISION POSITION ANALYTICAL SYNTHESIS*

In this section, four points were considered for the tool trajectory as mentioned in the above section (Table 2.5).

TABLE 2.5 Desired Four Precision Points for Tool Trajectory [18]

Desired Points (mm)	1	2	3	4
Pxd	125.116	141.741	180.804	194.196
Pyd	31.673	34.756	33.964	28.036

The position vector for the given prescribed coordinates for position 1, 2, 3 are written as $P_{\delta j} = R_1 - R_j$

The prescribed values for the following are:

$R_1 = 126.116 + 31.673i$; $R_2 = 141.741 + 34.756i$; $R_3 = 180.804 + 33.964i$;

$R_4 = 190.196 + 28.0361i$;

$P_{21} = -15.625 - 3.083i$; $P_{31} = -54.688 - 2.291i$; $P_{41} = -64.08 + 3.637i$

Using the theory of four precision position analytical synthesis methods, the solution is obtained. Figure 2.15 is the actual trajectory and through which the mechanism designed should pass through the selected precision point as explained in Section 2.3.2. There are infinite solutions in four precision position theory but all the solutions are not feasible. In the synthesis procedure, it is up to the designer to select which type of four-bar mechanism he needs for the prescribed path selected. Thus, he can take the option to consider more precision points through optimal synthesis path generation procedure for four-bar linkage. Four bar mechanism synthesized is shown in Figures 2.14 and 2.16 show that the mechanism traces the four precision point's path exactly.

$AD = r_1 = 570.735$ mm;
$AB = r_2 = 280.867$ mm;
$BC = r_3 = 450.980$ mm;
$CD = r_4 = 390.702$ mm.

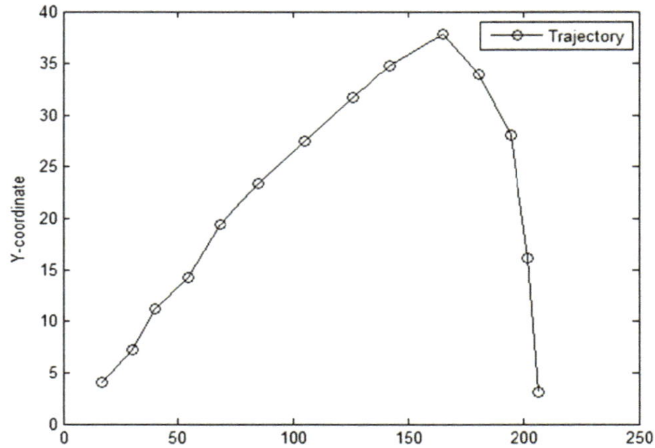

FIGURE 2.15 Actual trajectory and prescribed precision points.

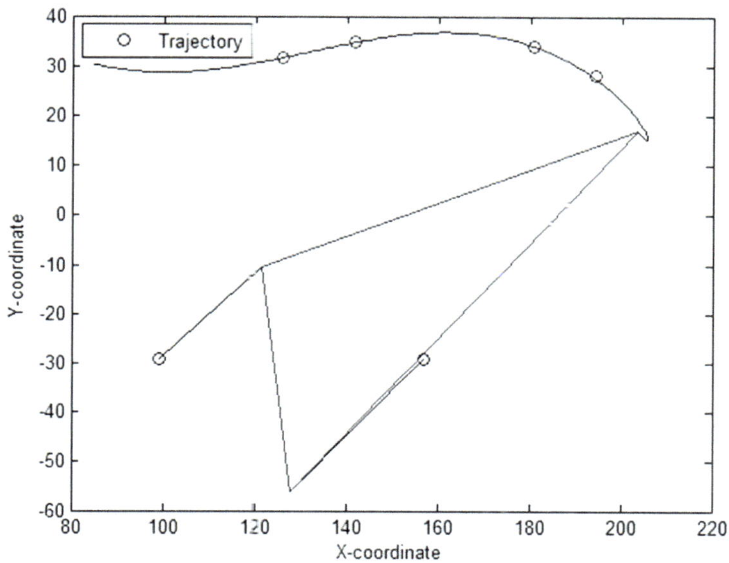

FIGURE 2.16 Synthesized four-bar linkage mechanism for four precision points using the analytical method.

2.3 CONCLUSION

The mechanism is designed for the vibratory tillage operation in this chapter. Three and four precision position analytical and graphical synthesis techniques are used to identify the dimensions, and it is confirmed that the dimensions obtained through graphical and analytical synthesis techniques are the same. Through MATLAB, also it is confirmed. Thus, the designed four-bar mechanism can be used for providing the oscillation to the tillage tool. According to the stability of the mechanism mounted on the cultivator frame, the dimensions can be selected obtained through the synthesis procedure.

ACKNOWLEDGMENT

The PhD scholarship is granted by the Ministry of Education, Government of India to the first author is highly acknowledged.

KEYWORDS

- **analytical linkage synthesis**
- **analytical method**
- **arbitrary precision points**
- **graphical synthesis**
- **precision position analytical synthesis**
- **tool trajectory**

REFERENCES

1. Report, (2016). *State of Indian Agriculture*. Ministry of Agriculture & Farmers Welfare, New Delhi, India.
2. Report, (2014). *Trends of Agricultural Mechanization in India CSAM Policy Brief* (Vol. 2, pp. 3–13). United Nations.
3. Mehta, C. R., Chandel, N. S., Jena, P. C., & Jha, A., (2019). Indian Agriculture counting on farm mechanization. *AMA-Agricultural Mechanization in Asia Africa and Latin America*, *50*(1), 84–89.
4. Bernacki, H., Haman, J., & Kanafojski, C., (1972). *Agricultural Machines: Theory and Construction* (Vol. 2). Washington, DC US: Department of Agriculture and the National Science Foundation.
5. Gunn, J. T., & Tramontini, V. N., (1955). Oscillation of tillage implements. *Agricultural Engineering*, *36*(11), 725–729.
6. Hendrick, J. G., & Buchele, W. F., (1963). Tillage energy of a vibrating tillage tool. *Transactions of the ASAE*, *6*(3), 213.
7. Johnson, C. E., & Buchele, W. F., (1969). Energy in the clod-size reduction of vibratory tillage. *Transactions of the ASAE*, *12*(3), 371.
8. Gupta, C. P., & Rajput, D. S., (1993). Effect of amplitude and frequency on soil break-up by an oscillating tillage tool in a soil bin experiment. *Soil and Tillage Research*, *25*(4), 329–338.
9. Niyamapa, T., & Salokhe, V. M., (2000). Soil disturbance and force mechanics of vibrating tillage tool. *Journal of Terramechanics*, *37*(3), 151–166.
10. Shahgoli, G., Saunders, C., Desbiolles, J., & Fielke, J., (2009). The effect of the oscillation angle on the performance of oscillatory tillage. *Soil and Tillage Research*, *104*(1), 97–105.
11. Razzaghi, E., & Sohrabi, Y., (2016). Vibratory soil cutting a new approach for the mathematical analysis. *Soil and Tillage Research*, *159*, 33–40.
12. Cooke, J. R., & Rand, R. H., (1969). Vibratory fruit harvesting: A linear theory of fruit-stem dynamics. *Journal of Agricultural Engineering Research*, *14*(3), 195–209.
13. Kofoed, S. S., (1969). Kinematics and power requirement of oscillating tillage tools. *Journal of Agricultural Engineering Research*, *14*(1), 54–73.

14. Sakai, K., Hata, S. I., Takai, M., & Nambu, S., (1993). Design parameters of four-shank vibrating subsoiler. *Transactions of the ASAE, 36*(1), 23–26.
15. Gupta, C. P., & Rajput, D. S., (1993). Effect of amplitude and frequency on soil break-up by an oscillating tillage tool in a soil bin experiment. *Soil and Tillage Research, 25*(4), 329–338.
16. Harrison, H. P., (1973). Draft, torque, and power requirements of a simple-vibratory tillage tool. *Canadian Agricultural Engineering, 15*(2), 71–74.
17. Russell, K., Shen, Q., & Sodhi, R. S., (2013). *Mechanism Design: Visual and Programmable Approaches*. CRC Press.
18. Shahgoli, G., Fielke, J., Desbiolles, J., & Saunders, C. (2010). Optimizing oscillation frequency in oscillatory tillage. *Soil and Tillage Research, 106*(2), 202–210.
19. Erdam, A. G., & Sandor, G. N., (1998). *Mechanism Design, Analysis, and Synthesis*, 291–353.
20. Upadhyay, G., & Raheman, H., (2019). Comparative analysis of tillage in sandy clay loam soil by free rolling and powered disc harrow. *Engineering in Agriculture, Environment, and Food, 12*(1), 118–125.
21. Rao, G., Chaudhary, H., & Singh, P., (2018). Optimal draft requirement for vibratory tillage equipment using a genetic algorithm technique. In: *IOP Conference Series: Materials Science and Engineering* (Vol. 330, No. 1, p. 012108). IOP Publishing.
22. Rao, G., & Chaudhary, H., (2018). A review on the effect of vibration in tillage application. In: *IOP Conference Series: Materials Science and Engineering* (Vol. 377, No. 1, p. 012030). IOP Publishing.
23. Rao, N. G., Chaudhary, H., & Sharma, A. K., (2019). Kinematic analysis of bionic vibratory tillage subsoiler. In: *Advances in Engineering Design* (pp. 187–195). Singapore: Springer.
24. Rao, G., Chaudhary, H., & Sharma, A., (2018). Design and analysis of vibratory mechanism for tillage application. *Open Agriculture, 3*(1), 437–443.
25. Rao, N. G., Chaudhary, H., & Sharma, A. K., (2019). Optimal design and analysis of oscillatory mechanism for agricultural tillage operation. *SN Applied Sciences, 1*(9), 1003.
26. Rao, G., Mall, N. K., Chaudhary, H., & Kumar, A., (2019). *Design of Four-Bar Mechanism for Transplanting Paddy Seedlings*. Available at: SSRN 3351776.
27. Mahmood, Y., Rao, G., Singh, P., & Chaudhary, H., (2019). Design modification for anti-choking mechanism in the thresher machine. In: *Machines, Mechanism, and Robotics* (pp. 585–593). Singapore: Springer.
28. Prem, M., Swarnkar, R., Kantilal, V. D. K., Jeetsinh, P. S. K., & Chitharbhai, K. B., (2016). Combined tillage tools: A review. *Current Agriculture Research Journal, 4*(2), 179–185.
29. Fenyvesi, L., & Hudoba, Z., (2010). Vibrating tillage tools. In: *Soil Engineering* (pp. 31–49). Berlin, Heidelberg: Springer.

CHAPTER 3

Mechanics of Metal Removal in Abrasive Jet Machining

V. DHINAKARAN,[1] JITENDRA KUMAR KATIYAR,[2] and T. JAGADEESHA[3]

[1]*Center for Applied Research, Chennai Institute of Technology, Chennai, Tamil Nadu, India*

[2]*SRM Institute of Science and Technology, Chennai, Tamil Nadu, India*

[3]*Department of Mechanical Engineering, NIT Calicut, India, E-mail: jagdishsg@nitc.ac.in*

ABSTRACT

In abrasive jet machining (AJM), a focused stream of abrasive particles, carried by high-pressure air or gas is made to impinge on the work surface through a nozzle, and work material is removed by erosion by high-velocity abrasive particles. The AJM differs from sandblasting in that the abrasive is much finer and the process parameters and cutting action are carefully controlled. AJM is mainly used to cut intricate shapes in a hard and brittle material, which are sensitive to heat and chip easily. The process is also used for deburring and cleaning operations. AJM is inherently free from chatter and vibration problems. The cutting action is cool because the carrier gas serves as a coolant. The high-velocity stream of abrasives is generated by converting the pressure energy of carrier gas or air to its kinetic energy and hence high-velocity jet. AJM consists of the gas propulsion system, abrasive feeder, machining chamber, AJM nozzle, and abrasives. Aluminum oxide (Al_2O_3), silicon carbide (SiC), glass beads, crushed glass, and sodium bicarbonate are some of the abrasives used in AJM. The selection of abrasives depends on MRR, type of work material, machining accuracy. This chapter deals with process details, experimental setup, process parameters used in AJM. The mechanics of metal removal in AJM is presented in this chapter. Mathematics of material removal models for ductile and brittle materials is discussed. Finally, practical examples are given to enhance the understanding of the mechanics of material removal in AJM.

3.1 INTRODUCTION

In abrasive jet machining, a focused stream of abrasive particles, carried by high-pressure air or gas is made to impinge on the work surface through a nozzle, and work material is removed by erosion by high-velocity abrasive particles. The AJM differs from sandblasting in that the abrasive is much finer and the process parameters and cutting action are carefully controlled [1–6].

Sandblasting	Abrasive Jet Machining
Sandblasting is an act of propelling very fine bits of material at high-velocity to clean or etch a surface.	In AJM, a focused steam of abrasive particles carried by high-pressure gas is used.
Uniform particles of sand, steel grit, copper slag, walnut shells, and powdered abrasives are used.	Silicon carbide, aluminum oxide, glass beads, dolomite, sodium bicarbonate are used as abrasives.
In sandblasting abrasive are sprayed all over usually for cleaning surfaces from corrosion, paints, glues, and other contaminants	AJM is not only used for cleaning but also used for cutting, deburring, deflashing, etc. AJM is a well-controlled process compared to sandblasting

AJM is mainly used to cut intricate shapes in hard and brittle material which are sensitive to heat and chip easily. The process is also used for deburring and cleaning operations. AJM is inherently free from chatter and vibration problems. The cutting action is cool because the carrier gas serves as a coolant.

3.2 DESCRIPTION OF PROCESS

The schematic diagram of abrasive jet machining is shown in Figure 3.1. In abrasive jet machining, abrasive particles are made to impinge on work material at high velocity. Jet of abrasive particles is carried by carrier gas or air. The high-velocity stream of abrasives is generated by converting the pressure energy of carrier gas or air to its kinetic energy and hence high-velocity jet. Nozzles direct abrasive jet in a controlled manner onto work material. The high-velocity abrasive particles remove the material by micro-cutting action as well as brittle fracture of the work material.

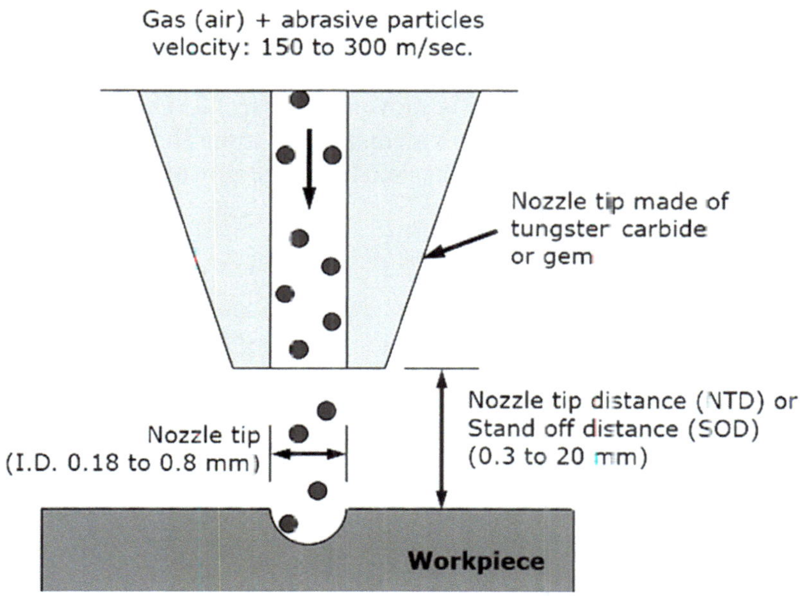

FIGURE 3.1 Abrasive jet process.

This is a process of removal of material by impact erosion through the action of a concentrated high-velocity stream of grit abrasives entrained in the high-velocity gas stream. AJM is different from shot or sandblasting, as in AJM, finer abrasive grits are used and parameters can be controlled more effectively providing better control over product quality.

In AJM, generally, the abrasive particles of around 50 microns grit size would impinge on the work material at a velocity of 200 m/s from a nozzle of ID 0.5 mm with a standoff distance of around 2 mm. The kinetic energy of the abrasive particles would sufficient to provide material removal due to brittle fracture of the workpiece or even micro-cutting by the abrasives. The four important actions in abrasive jet machining are [7–12]:

- Fine particles (0.025 mm) are accelerated in a gas stream.
- The particle is directed towards the focus of machining.
- As the particles impact the surface, it causes a microfracture, and gas carries fractured particles away.
- Brittle and fragile work better.

3.3 DESCRIPTION OF EQUIPMENT

A schematic layout of the AJM is shown in Figure 3.2. The gas stream is then passed to the nozzle through a connecting hose. The velocity of the abrasive stream ejected through the nozzle is generally of the order of 330 m/sec.

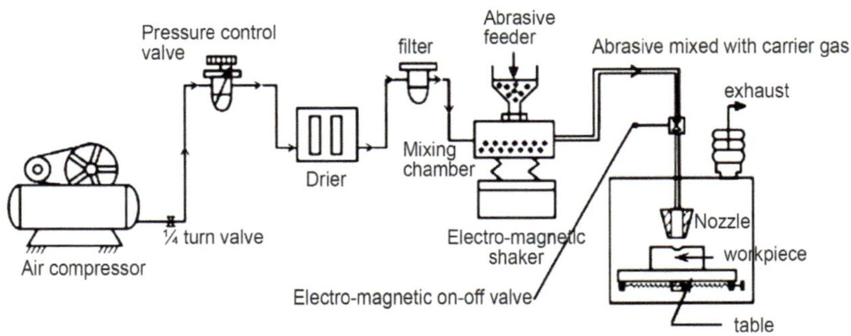

FIGURE 3.2 Schematics of AJM.

Abrasive jet machining consists of:

1. Gas propulsion system;
2. Abrasive feeder;
3. Machining chamber;
4. AJM nozzle;
5. Abrasives.

3.3.1 GAS PROPULSION SYSTEM

Gas propulsion system supplies clean and dry air. Air, nitrogen, and carbon dioxide propel the abrasive particles. Gas may be supplied either from a compressor or a cylinder. In the case of a compressor, an air filter cum drier should be used to avoid water or oil contamination of abrasive powder. Gas should be non-toxic, cheap, and easily available. It should not excessively spread when discharged from the nozzle into the atmosphere. The propellant consumption is of the order of 0.008 m^3/min at a nozzle pressure of 5 bar and the abrasive flow rate varies from 2 to 4 g/min for fine machining and 10 to 20 g/min for cutting operation [13–17]

3.3.2 ABRASIVE FEEDER

The required quantity of abrasive particles is supplied by an abrasive feeder. The filleted propellant is fed into the mixing chamber wherein abrasive particles are fed through a sieve. The sieve is made to vibrate at 50–60 Hz and the mixing ratio is controlled by the amplitude of vibration of the sieve. The particles are propelled by a carrier gas (transfer medium) to a mixing chamber. Air abrasive mixture moves further to the nozzle. The nozzle imparts high velocity to the mixture which is directed at the workpiece surface.

The desired properties of carrier gas used in AJM are:

1. It should be non-toxic;
2. It should be cheap and easily available;
3. It should not spread too much when discharged from the nozzle to the atmosphere;
4. It should be easy to condition the carrier gas before mixing with abrasives: like drying to remove moisture, filtering to remove the contaminations.

Nitrogen, carbon dioxide, helium, and compressed air are commonly used carrier gas in AJM.

3.3.3 MACHINING CHAMBER

It is well closed so that the concentration of abrasive particles around the working chamber does not reach to the harmful limits. The machining chamber is equipped with a vacuum dust collector. Special consideration should be given to the dust collection system if the toxic material (like beryllium) is being machined.

3.3.4 AJM NOZZLE

AJM nozzle is usually made of tungsten carbide or sapphire (usually life-300 hours for sapphire, 20 to 30 hours for WC) which has resistance to wear. The nozzle is made of either circular or rectangular cross-section and the head can be straight or at a right angle. It is so designed that loss of pressure due to the bends, friction, etc., is minimum possible. With an increase in wear of a nozzle, the divergence of jet stream increases resulting in more stray cutting and high inaccuracy.

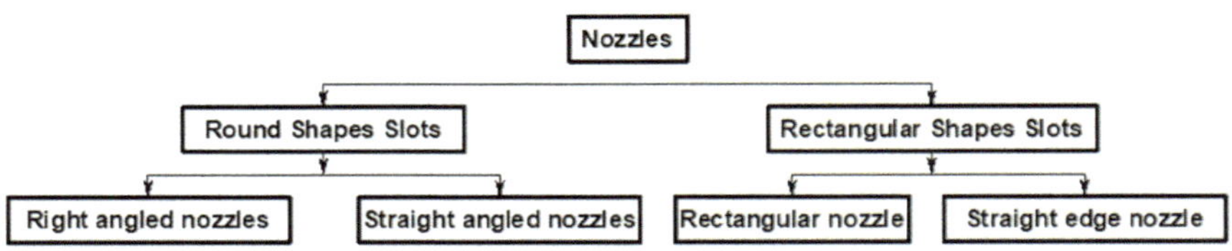

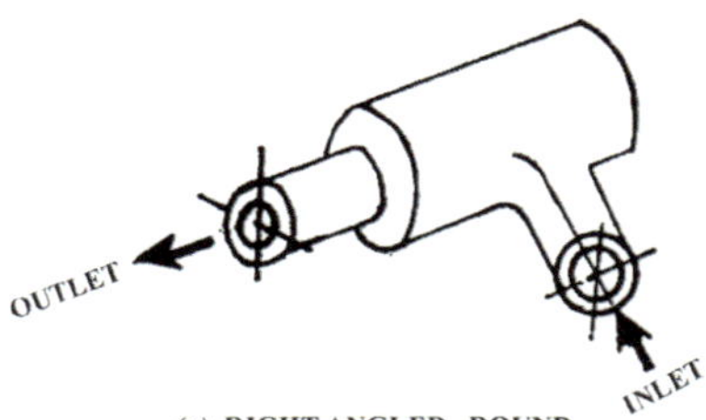

(a) RIGHT-ANGLED, ROUND SHAPED NOZZLE

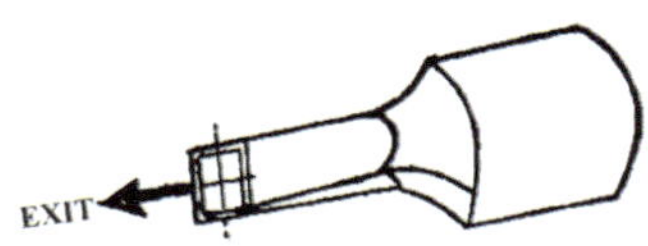

(b) STRAIGHT EDGE RECTANGULAR SHAPED NOZZLE

Nozzle Material	Round Shape Nozzle, Diameter (mm)	Rectangular Shape Slot, Dimension (mm)	Life of Nozzle (Hours)
Tungsten Carbide (WC)	0.2 to 1.0	0.075×0.5 to 0.15×2.5	12 to 30
Sapphire	0.2 to 0.8	–	300

The desired properties of materials used for the nozzle are:

1. Material has to withstand the erosive action of abrasive particles.
2. It should have good wear resistance properties: Increase in the wear of the nozzle leads to the divergence of the jet stream. Divergence of jet steam causes stray cutting and inaccurate holes.
3. It should have good resistance to corrosion.
4. It should be designed such that loss of pressure due to bend and friction is minimum.

3.3.5 ABRASIVES USED

Aluminum oxide (Al_2O_3), silicon carbide (SiC), glass beads, crushed glass, and sodium bicarbonate are some of the abrasives used in AJM. The selection of abrasives depends on MRR, type of work material, machining accuracy. Table 3.1 gives the classification of abrasives and their applications.

TABLE 3.1 Different Abrasives Used in AJM and Its Applications

Abrasives	Grain Sizes	Application
Aluminum oxide (Al_2O_3)	12, 20, 50 microns	Good for cleaning, cutting, and deburring
Silicon carbide (SiC)	25, 40 microns	Used for similar application but for hard material
Glass beads	0.635 to 1.27 mm	Gives matte finish
Dolomite	200 mesh	Etching and polishing
Sodium bicarbonate	27 micros	Cleaning, deburring, and cutting of soft material Light finishing below 50°C

The desired properties of abrasives used in AJM are:

1. Abrasives should have a sharp and irregular shape.
2. It should be fine enough to remain suspended in a carrier gas.
3. In addition to hardness, the important properties of abrasive are friability. Friability is the ability of abrasive grains to fracture

into smaller pieces; this property gives abrasives self-sharpening characteristics. High friability indicates low strength or low fracture resistance of the abrasives. High friable abrasives fragment more rapidly than lower friable abrasives. Aluminum oxide has a lower friability than silicon carbide; hence, it has less tendency to fracture.

4. Should be cheap and easily available.
5. It should have excellent flow characteristics.

Silicon carbide and aluminum oxide are used for cutting operations. Sodium bicarbonate, dolomite, and glass beads are used for cleaning, etching, deburring, and polishing applications.

The reuse of abrasives is not recommended because:

1. Cutting capacity decreases after the first use.
2. Contamination clogs the small orifices in the nozzle.

3.4 PROCESS PARAMETERS

For the successful utilization of the AJM process, it is necessary to analyze the following process criteria:

1. Material removal rate.
2. Geometry and surface. Finish of workpiece.
3. Wear rate of the nozzle.

However, process criteria are generally influenced by the process parameters as enumerated below:

- **Abrasives:**
  - **Material:** Al_2O_3SiC, Glass beads, Crushed glass, Sodium bicarbonate;
  - **Shape:** Irregular/regular;
  - **Size:** 10 to 50 microns;
  - **Mass flow:** 2–20 grams/min.
- **Carrier Gas:**
  - **Composition:** Air, CO_2, N_2;
  - **Density:** 1.3 kg/m^3;
  - **Velocity:** 500 to 700 m/s;

  - **Pressure:** 2 to 10 bars;
  - **Flow rate:** 5 to 30 microns.
- **Abrasive Jet:**
  - **Velocity:** 100 to 300 m/s;
  - **Mixing ratio:** Volume flow rate of abrasives/Volume flow rate of gas;
  - **Standoff distance (SOD):** 0.5 to 15 mm;
  - **Impingement angle:** 60 to 90 deg.
- **Nozzle:**
  - **Material:** WC/Sapphire;
  - **Diameter:** 0.2 to 0.8 mm;
  - **Life:** 300 hours for sapphire, 20 to 30 hours for WC.

3.5 PROCESS CAPABILITY

1. **Material removal rate:** 0.015 cm^3/min
2. **Narrow slots:** 0.12 to 0.25 mm ± 0.12 mm
3. **Surface finish:** 0.25 micron to 1.25 micron
4. **Sharp radius:** up to 0.2 mm is possible
5. Steel up to 1.5 mm, glass up to 6.3 mm is possible to cut
6. Machining of thin sectioned hard and brittle materials is possible.

3.6 APPLICATIONS OF AJM

1. This is used for abrading and frosting glass, ceramics, and refractories more economically as compared to etching or grinding.
2. Cleaning of metallic smears on ceramics, oxides on metals, resistive coating, etc.
3. AJM is useful in the manufacture of electronic devices, drilling of glass wafers, deburring of plastics, making of nylon and Teflon parts permanent marking on rubber stencils, cutting titanium foils.
4. Deflashing small castings and trimming of parting lines of injection molded parts and forgings.
5. Use for engraving registration numbers on toughened glass used for car windows.
6. Used for cutting thin fragile components like germanium, silicon, quartz, mica, etc.

7. Register streaming can be done very easily and micro module fabrication for electrical contact, semiconductor processing can also be done effectively.
8. Used for drilling, cutting, deburring etching, and polishing of hard and brittle materials.
9. Most suitable for machining brittle and heat-sensitive materials like glass, quartz, sapphire, mica, ceramics germanium, silicon, and gallium.
10. It is also a good method for deburring small holes like in hypodermic needles and for small milled slots in hard metallic components.
11. It can be used for micromachining of brittle materials.
12. It is used in fine drilling and aperture drilling for an electronic microscope.
13. Used for cleaning metallic molds and cavities.
14. Cleaning surfaces from corrosion, paints, glues, and other contaminants, especially those that are inaccessible.
15. Deburring of surgical needles and hydraulic valves, nylon, teflon, and derlin.
16. Engraving on glass using rubber or metallic masks.

3.7 ADVANTAGES OF AJM

1. High surface finish can be obtained depending upon the grain sizes

Particle Size (in Microns)	Surface Roughness (in Microns)
10	0.152 to 0.203
25 to 27	0.355 to 0.675
50	0.965 to 1.27

2. Depth of damage is low (around 2.5 microns).
3. It provides cool cutting action, so it can machine delicate and heat-sensitive material such as glass and ceramics. They can be machined without affecting their physical properties and crystalline structure.
4. Process is free from chatter and vibration as there is no contact between the tool and the workpiece
5. Capital cost is low and it is easy to operate and maintain AJM.
6. Thin sections of hard brittle materials like germanium, mica, silicon, glass, and ceramics can be machined.

7. It has the capability of cutting holes of intricate shape in hard and brittle materials.
8. Abrasive jet processes produce surfaces which have high wear resistance.

3.8 DISADVANTAGES OF AJM

1. Limited capacity due to low MRR. MRR for glass is 40 g/min.
2. Abrasives may get embedded in the work surface, especially while machining soft material like elastomers or soft plastics.
3. The accuracy of cutting is hampered by tapering of the hole due to the unavoidable flaring of the abrasive jet.
4. Stray cutting is difficult to avoid and hence accuracy is not good.
5. A dust collection system is a basic requirement to prevent atmospheric pollution and health hazards.
6. Nozzle life is limited (300 hours).
7. Abrasive powders cannot be reused as the sharp edges are worn and smaller particles can clog the nozzle.
8. Short standoff distances when used for cutting, damages the nozzle.
9. The process accuracy is poor because of the flaring effect of the abrasive jet.
10. Deep holes will have an unacceptable taper.
11. Process is not environmentally friendly and causes pollution.
12. Some hazard is involved in using AJM process because of airborne abrasives particulates. By using abrasive water jet machining this problem can be solved.

3.9 MACHINING CHARACTERISTICS OF AJM

The following are the AJM process criteria:

1. Material removal rate;
2. Geometry and surface finish of workpiece;
3. Wear rate of the nozzle.

Process criteria are generally influenced by the process parameters The characteristics of the above process parameters on process criteria are as follows:

1. **Effect of Abrasive Flow Rate and Grain Size on MRR (Figure 3.3):** It is clear from the figure that at a particular pressure MRR increases with an increase of abrasive flow rate and is influenced by the size of abrasive particles. But after reaching optimum value, MRR decreases with a further increase of abrasive flow rate. This is owing to the fact that Mass flow rate of the gas decreases with an increase of abrasive flow rate and hence mixing ratio increases causing a decrease in material removal rate because of decreasing energy available for erosion.

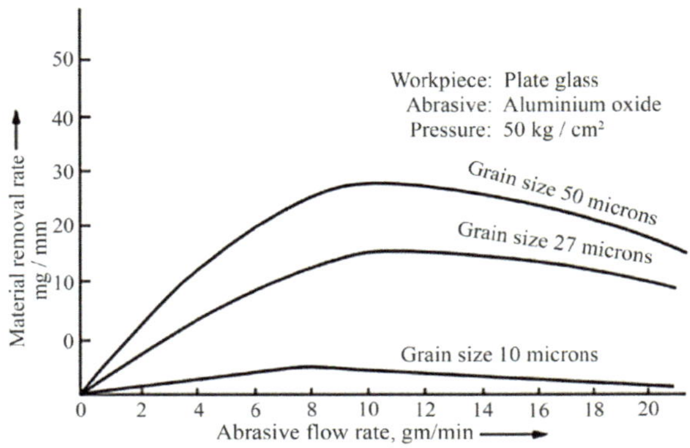

FIGURE 3.3 Effect of abrasive flow rate and grain size on MRR.

2. **Effect of Exit Gas Velocity and Abrasive Particle Density (Figure 3.4):** The velocity of carrier gas conveying the abrasive particles changes considerably with the change of abrasive particle density as indicated in Figure 3.4.

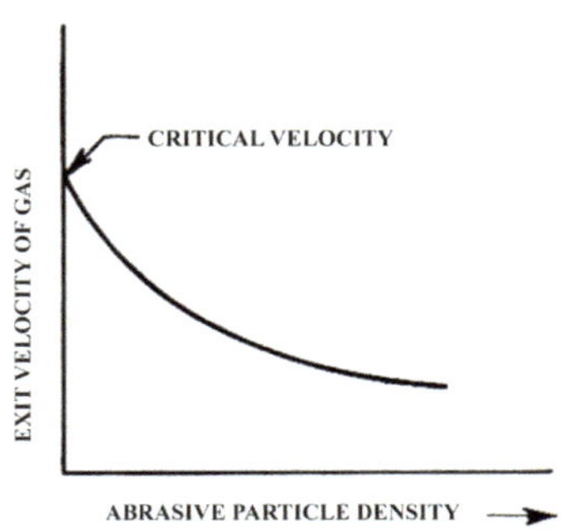

FIGURE 3.4 Effect of exit gas velocity and abrasive particle density.

The exit velocity of gas can be increased to critical velocity when the internal gas pressure is near twice the pressure at the exit of the nozzle for the abrasive particle density is zero. If the density of abrasive particles is gradually increased exit velocity will go on decreasing for the same pressure condition. It is due to the fact that the kinetic energy of the gas is utilized for transporting the abrasive particle

3. **Effect of Mixing Ratio on MRR:** Increased mass flow rate of abrasive will result in a decreased velocity of the fluid and will thereby decrease the available energy for erosion and ultimately the MRR. It is convenient to explain to this fact by the term *mixing ratio*, which is defined as

$$Mixing\ ratio = \frac{Volume\ flow\ rate\ of\ abrasives\ per\ unit\ time}{Volume\ flow\ rate\ of\ carrier\ gas\ per\ unit\ time}$$

The effect of the mixing ratio on the material removal rate is shown in Figure 3.5.

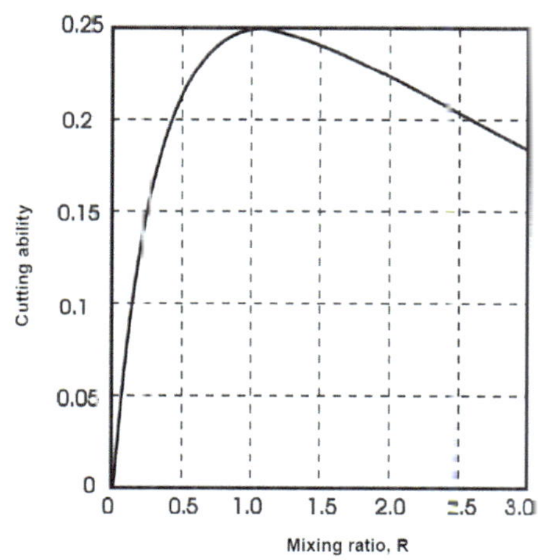

FIGURE 3.5 Effect of mixing ratio on MRR.

The material removal rate can be improved by increasing the abrasive flow rate provided the mixing ratio can be kept constant. The mixing ratio is unchanged only by the simultaneous increase of both gas and abrasive flow rate (Figure 3.6).

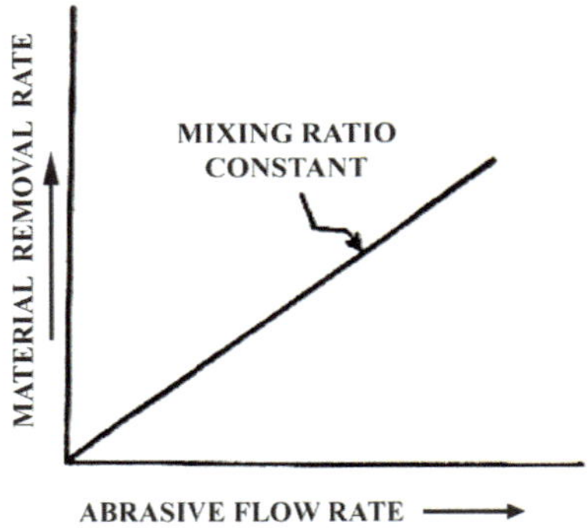

FIGURE 3.6 Effect of abrasive flow on MRR with constant MR.

An optimum value of mixing ratio that gives maximum MRR is predicted by trial and error. In place of the mixing ratio, the mass ratio α may be easier to determine, which is defined as:

$$\alpha = \frac{Volume\ flowrate\ of\ abrasives}{Volume\ flowrate\ of\ carrier\ gas} = \frac{m_a}{m_a + m_g}$$

4. **Effect of Nozzle Pressure on MRR:** The abrasive flow rate can be increased by increasing the flow rate of the carrier gas. This is only possible by increasing the internal gas pressure as shown in Figure 3.7. As the internal gas pressure increases abrasive mass flow rate increases and thus MRR increases.

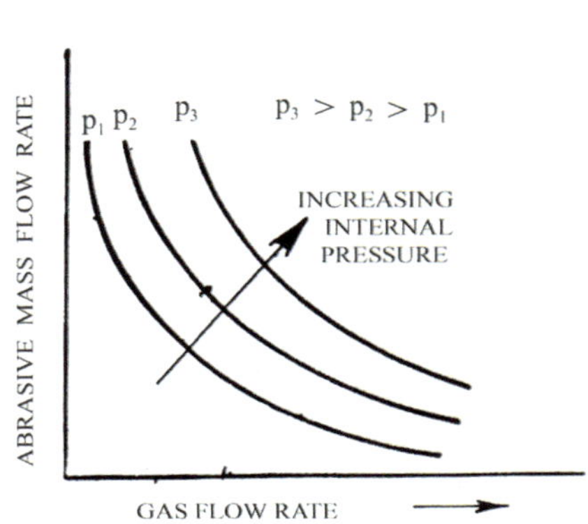

FIGURE 3.7 Effect of gas flow rate on abrasive mass flow rate with pressure.

As a matter of fact, the material removal rate will increase with the increase in gas pressure

The kinetic energy of the abrasive particles is responsible for the removal of material by the erosion process. The abrasive must impinge on the work surface with minimum velocity for machining glass by SIC particle is found to be around 150 m/s (Figure 3.8).

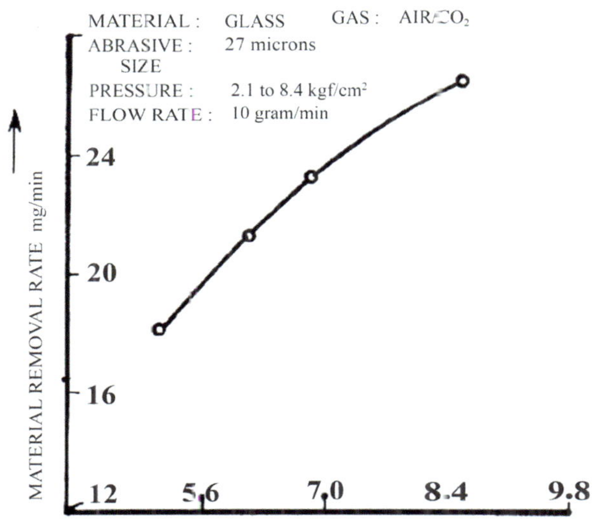

FIGURE 3.8 Effect of gas flow rate on MRR.

5. **Standoff Distance:** It is defined as the distance between the face of the nozzle and the work surface of the work. SOD has been found to have a considerable effect on the work material and accuracy. A large SOD results in flaring of the jet which leads to poor accuracy.

It is clear from Figure 3.9 that MRR increases with nozzle tip distance or Standoff distance up to a certain distance and then decreases. Penetration rate also increases with SOD and then decreases. Decrease in SOD improves accuracy, decreases kerf width, and reduces taper in the machined groove. However, light operations like clearing, frosting, etc., are conducted with large SOD (say 12.5 to 75 mm) (Figure 3.10).

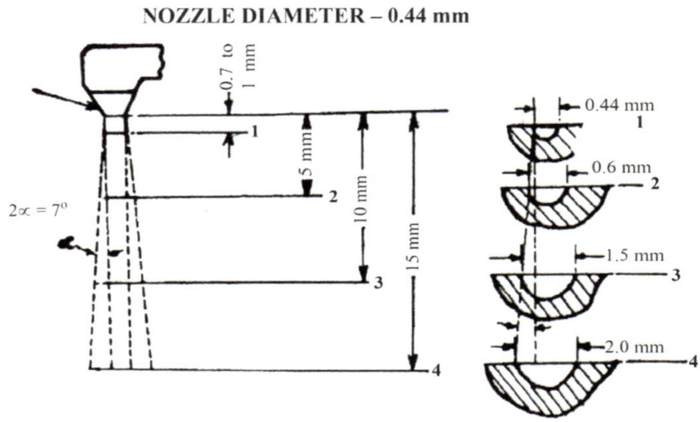

FIGURE 3.9 Effect of standoff distance on the width of the cut.

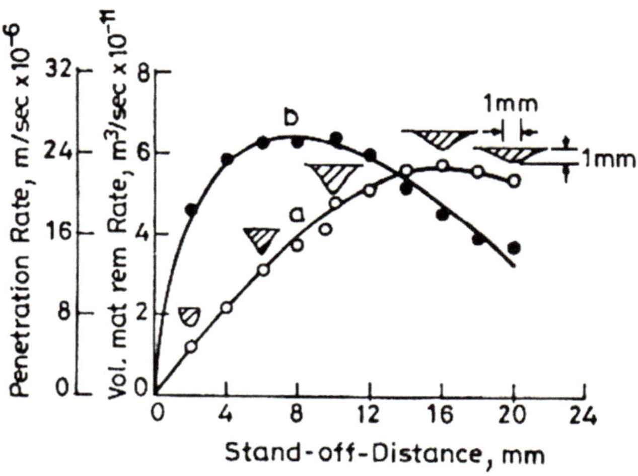

FIGURE 3.10 Effect of standoff distance on the material removal rate and penetration rate.

3.10 MECHANICS OF CUTTING IN AJM

3.10.1 MATERIAL REMOVAL MODELS IN AJM

The following assumptions are made in deriving the material removal models for AJM:

1. Abrasive is spherical in shape and rigid.
2. Kinetic energy of the particle is completely used to cut the material.
3. Brittle materials are considered to fail due to brittle fracture and fracture of the volume is considered to be hemispherical with a diameter equal to the chordal length of indentation.
4. For the Ductile material volume of material, removal is assumed to be equal to indentation volume due to particulate impact.

The abrasive particles are assumed to be spherical in shape having a diameter d_g. From the geometry (Figure 3.11):

$$AB^2 = AC^2 + BC^2$$

$$\frac{d_g}{2} = \left(\frac{d_g}{2} - \delta\right)^2 + r^2$$

$$\frac{d_g}{2} - \left(\frac{d_g}{2} - \delta\right)^2 = r^2$$

$$r^2 = \delta^2 + d_g\delta$$

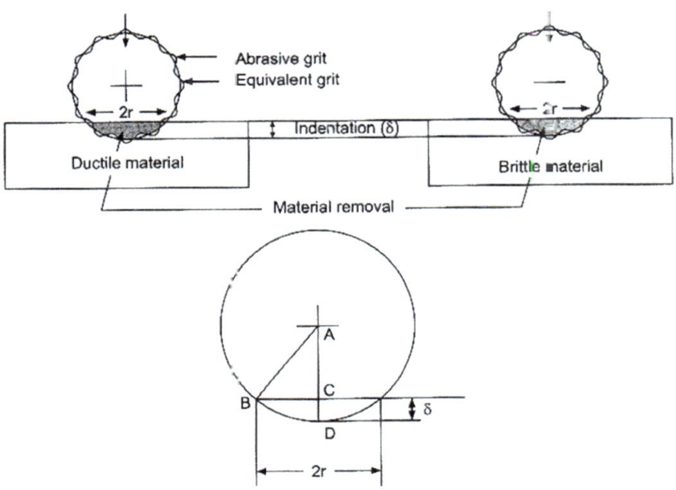

FIGURE 3.11 Material removal models for AJM.

Neglecting the δ^2 term we can write:

$$r^2 = d_g\delta$$

$$r = \sqrt{d_g\delta}$$

3.10.2 FOR BRITTLE MATERIAL

The volume of the material removed is the volume of the hemispherical crater due to the fracture is given by

$$\Gamma_B = \frac{1}{2}\left[\frac{4}{3}\pi r^3\right] = \frac{1}{2}\left[\frac{4}{3}\pi\left(r^2\right)^{\frac{3}{2}}\right] = \frac{1}{2}\left[\frac{4}{3}\pi\left(d_g\delta\right)^{\frac{3}{2}}\right] = \frac{2\pi}{3}\left(d_g\delta\right)^{\frac{3}{2}} \tag{1}$$

Let us assume that grits also move with velocity (V) then we can write,

Kinetic energy = KE = $\frac{1}{2}MV^2$

where, M = mass of a single abrasive grit = volume of grit × density of grit.

$$Volume\ of\ single\ grit = \frac{4\pi r_g^{\ 3}}{3} = \frac{4\pi\left(\frac{d_g}{2}\right)^3}{3} = \frac{\pi\left(d_g\right)^3}{6}$$

Therefore, we can write kinetic energy of the single grit

$$KE = \frac{1}{2}MV^2 = \frac{1}{2}\left(\frac{\pi\left(d_g\right)^3}{6}\rho_g\right)V^2 \tag{2}$$

On impact, work material would be subjected to maximum force F, which would lead to indentation of δ. The work done (WD) during such indentation is:

$$\text{WD by the grit} = \frac{F\delta}{2} \tag{3}$$

Also, we know the flow strength of material = $\sigma_w = \frac{F}{\pi r^2}$

$$F = \sigma_w \times \pi r^2$$
$$F = \sigma_w \times \pi d_g \delta \qquad (4)$$

Using Eq. (4) in (3) we get:

WD by the grit = $\dfrac{F\delta}{2} = \dfrac{\sigma_w \times \pi d_g \delta \times \delta}{2}$

It is assumed that the Kinetic energy of the abrasives is fully used for material removal.

Kinetic energy of the particle = WD by the particle:

$$\frac{1}{2}\left(\frac{\pi (d_g)^3}{6}\rho_g\right)V^2 = \frac{\sigma_w \times \pi d_g \delta \times \delta}{2}$$

Simplifying we get:

$$\delta = V d_g \sqrt{\frac{\rho_g}{6\sigma_w}} \qquad (5)$$

MRR in AJM material can be expressed as:

$$MRR = \{Volume\ of\ material\ removed\ per\ grit\ per\ cycle\} \times Number\ of\ impacts\ made\ by\ abrasives\ per\ second$$

$$MRR = \Gamma_B \times \left[\frac{Mass\ flow\ rate\ of\ abrasives}{Mass\ of\ the\ abrasive\ grit}\right]$$

$$MRR = \frac{2\pi}{3}(d_g \delta)^{\frac{3}{2}} \times \left[\frac{M_a}{\dfrac{\pi (d_g)^3}{6}\rho_g}\right]$$

Upon simplifying we get:

$$MRR = \left[\frac{M_a (V)^{\frac{3}{2}}}{(\sigma_w)^{\frac{3}{4}} (\rho_g)^{\frac{1}{4}}} \right]$$

3.10.3 FOR DUCTILE MATERIAL

The volume of the material removed in a single impact is equal to the volume of indentation.

$$\Gamma_D = \pi \delta^2 \left[\frac{d_g}{2} - \frac{\delta}{3} \right] = \frac{\pi \delta^2 d_g}{2}$$

MRR in AJM material can be expressed as:

$$MRR = \{Volume\ of\ material\ removed\ per\ grit\ per\ cycle\} \times Number\ of\ impacts\ made\ by\ abrasives\ per\ second$$

$$MRR = \Gamma_D \times \left[\frac{Mass\ flow\ rate\ of\ abrasives}{Mass\ of\ the\ abrasive\ grit} \right]$$

$$MRR = \frac{\pi \sigma^2 d_g}{w} \times \left[\frac{M_a}{\frac{\pi (dg)^3}{6} \rho_g} \right]$$

Substituting and simplifying we get:

$$MRR = \frac{\pi \delta^2 d_g}{2} \times \left[\frac{M_a V^2}{2\sigma_w} \right]$$

3.11 WORKED EXAMPLES

- **Problem 1:** Estimate the MRR in AJM of a brittle material with a flow strength of 3 GPa. The abrasive flow rate is 2.5 g/min, velocity is 205 m/s, density of abrasive is 3 g/sec.

Data Given:

Flow strength of work material = $\sigma_w = 3\times10^9 \dfrac{N}{\text{m}^2}$

Abrasive grain density $= \rho_g = 3\dfrac{\text{grams}}{\text{CC}} = \dfrac{3\times10^{-3}}{10^{-6}}\dfrac{\text{kg}}{\text{m}^3} = 3\times10^3\dfrac{\text{kg}}{\text{m}^3}$

Mass flow rate of abrasives $= M_a = \dfrac{2.5\text{ grams}}{\text{min}} = \dfrac{2.5\times10^{-3}}{60}\text{kg/sec}$

$$Velocity = 205 \text{ m/sec}$$

Solution:
Since the material is brittle, we need to use the MRR formula corresponding to the brittle material.

$$MRR = \left[\frac{M_a (V)^{\frac{3}{2}}}{(\sigma_w)^{\frac{3}{4}} (\rho_g)^{\frac{1}{4}}}\right]$$

$$MRR = \left[\frac{2.5\times10^{-3}/60\times(205)^{\frac{3}{2}}}{\left(3\times10^9\right)^{\frac{3}{4}}\left(3\times10^3\right)^{\frac{1}{4}}}\right]$$

$$MRR = 1.289\times10^{-9}\frac{m^3}{\text{sec}} = 77.35\frac{mm^3}{\text{min}}$$

- **Problem 2:** The material removal rate in AJM is 0.5 mm³/sec. calculate MRR/impact if the mass flow rate of abrasive is 3 g/min, density is 3 g/CC and grit size is 60 microns. Also, calculate the indentation radius.

Data Given:
Material removal rate = 0.5 mm³/sec.
Abrasive grain size = 50 microns = 50×10^{-3} mm

Abrasive grain density $\rho_g = 3\dfrac{\text{grams}}{\text{CC}} = \dfrac{3\times10^{-3}}{10^{-6}}\dfrac{\text{kg}}{\text{m}^3} = 3\times10^3\dfrac{\text{kg}}{\text{m}^3}$

Mass flow rate of abrasives M_a = 3 g/min = $3\times10^{-3}/60$ kg/sec

Solution:

$$Volume\ of\ single\ grit = \frac{4\pi r_g^{\ 3}}{3} = \frac{4\pi\left(\frac{d_g}{2}\right)^3}{3} = \frac{\pi\left(d_g\right)^3}{6}$$

$$Mass\ of\ single\ grit = \frac{\pi\left(d_g\right)\rho_g^{\ 3}}{6} = \frac{\pi\left(50\times10^{-3}\right)\left(\frac{3\times10^{-3}}{60}\right)^3}{6}$$

MRR in AJM material can be expressed as:

$$MRR = \{Volume\ of\ material\ removed\ per\ grit\ per\ cycle\} \times Number\ of\ impacts\ made\ by\ abrasives\ per\ second$$

$$Number\ of\ impacts\ made\ by\ abrasives\ per\ second\ (N) = \frac{Mass\ flow\ rate\ of\ abrasives}{Mass\ of\ the\ abrasive\ grit}$$

$$N = \frac{M_a}{Mass\ of\ the\ abrasive\ grit} = \frac{\frac{3\times10^{-3}}{60}}{\frac{\pi\left(50\times10^{-3}\right)(3000)}{6}} = 254648\ impacts/s$$

$$0.5 = \{Volume\ of\ material\ removed\ per\ grit\ per\ cycle\} \times Number\ of\ impacts\ made\ by\ abrasives\ per\ second$$

$$0.5 = \{Volume\ of\ material\ removed\ per\ grit\ per\ cycle\} \times 254648$$

$$0.5 = \Gamma_B \times 254648$$

$$\Gamma_B = volume\ of\ material\ removed\ per\ grit = \left[\frac{0.5}{254648}\right] = 1.96\times10^{-6}\ mm^3 = 1960\ \mu m^3$$

$$Indentation\ Volume = \frac{1}{2}\left(\frac{4\pi r_g^{\ 3}}{3}\right) = \frac{1}{2}\left[\frac{4\pi\left(\frac{d_g}{2}\right)^3}{3}\right] = \frac{\pi\left(d_g\right)^3}{12} = 1960$$

Solving we get, d_g = 19.56 microns or indentation radius ≅ 10 microns.

- **Problem 3:** During AJM, the mixing ratio used is 0.2. Calculate mass ratio, if the ratio of the density of abrasives and density of carrier gas is equal to 20.

Solution:

$$Mixing\ ratio = \frac{Volume\ flow\ rate\ of\ abrasive\ particles}{Volume\ flow\ rate\ of\ carrier\ gas} = \frac{V_g}{V_a} = 0.2$$

$$Mixing\ ratio(\alpha) = \frac{Abrasive\ mass\ flow\ rate}{combined\ mass\ flow\ rate\ of\ abrasive\ and\ carrier\ gas}$$

$$Mass\ ratio(\alpha) = \frac{M_a}{M_a + M_g} = \frac{\rho_a V_a}{\rho_a V_a + \rho_g V_g}$$

We can rewrite the expression for mass ratio as:

$$\frac{1}{\alpha} = \frac{M_a + M_g}{M_a} = \frac{\rho_a V_a + \rho_g V_g}{\rho_a V_a} = 1 + \left[\frac{\rho_g}{\rho_a}\right]\left[\frac{V_g}{V_a}\right] = 1 + \left[\frac{1}{20}\right]\left[\frac{1}{0.2}\right]$$

Solving we get, $\alpha = 0.80$

- **Problem 4:** Diameter of the nozzle is 1.0 mm and jet velocity is 200 m/s. Find the volumetric flow rate in cm³/sec and in mm³/sec of carrier gas and abrasive mixture.

Solution:

Cross-sectional area of nozzle $= \pi \times 0.5^2 \times 10^{-2} cm^2$

Volumetric flow rate of carrier gas and the abrasive mixture is $\left(V_{gas+mixture}\right)$

$$\left(V_{gas+mixture}\right) = area \times Velocity = \pi \times 0.5^2 \times 10^{-2} \times 2$$

$$= 157.075\, cm^3 / \sec = 157075\, mm^3 / \sec$$

KEYWORDS

- **abrasive mass flow rate**
- **abrasive particle density**
- **exit gas velocity**
- **sandblasting**
- **standoff distance**

REFERENCES

1. Springeborn, R. K., (1967). *Nontraditional Machining Processes*. Michigan: ASTME.
2. Sheldon, G. L., & Finnie, T., (1980). The mechanics of metal removal in erosion cutting of brittle materials. *Tans. ASME Series B, 88*.
3. Lvoie, F. J., (1973). Abrasive jet machining. *Machine Design*.
4. LaCourte, N., (1979). Abrasive JET machining-a solution for problem jobs. *Tooling and Production*, 104–106.
5. Sarkar, P. K., & Pandey, P. C., (1975). *Some Investigation on the Abrasive Jet Machining*. Institution of India.
6. Arion, (1963). *Abrasive Machining*. ASTME, Book 6. Collected series.
7. Pandey, P. C., & Neema, M. L., (1977). *Erosion of Glass When Acted Upon by an Abrasive Jet* (pp. 387–391). St. Louis: International Conference on Wear of Materials.
8. Sapra, I. L., (1975). *Studies on Abrasive Jet Machining*. M. E. Dissertation, University of Roorkee.
9. Dombrowski, T. R., (1983). *The How and Why of Abrasive Je Machining* (pp. 76–77). Modern Machine Shop.
10. Finnie, I., (1966). The mechanism of metal removal in the erosion cutting of brittle materials. *Trans ASME, Ser, B., 88*, 393.
11. Ingulli, C. N., (1967). Abrasive jet machining, too and manuf. *Engrs., 59*, 28.
12. Sheldon, G. L., & Finnie, I., (1966). The mechanics of material removal in the erosive cutting of brittle materials. *Trans. ASME Ser B., 88*, 393.
13. Shoeyes, R., Stalens, F., & Dekeyser, W., (1986). Current trends in non-conventional machining processes. *Annals of CIRP, 35*(2), 467.
14. Venkatesh, V. C., (1984). parametric studies on Abrasive jet machining. *Annals of the CIRP, 33*(1), 109.
15. Verma, A. P., & Lal, G. K., (1985). Basic mechanics of abrasive jet machining. *International Journal of Engrs. Prod. Engg., 66*, 74–81.
16. Verma, A. P., & Lal, G. K., (1984). An experimental study of abrasives jet machining. *Int. J. Mach. Tools Des. Res., 24*(1), 19–29.
17. Bitter, J. G. A., (1962). A study of erosion phenomenon. *Wear, Part 1 and II, 5, 6, 6*, 169–190.

CHAPTER 4

Evaluating the Mass Sensing Characteristics of SWCNC

UMANG B. JANI, BHAVIK A. ARDESHANA, AJAY M. PATEL, and ANAND Y. JOSHI

Mechatronics Engineering Department, G. H. Patel College of Engineering & Technology, Vallabh Vidyanagar, Gujarat, India, E-mails: umangjani@gcet.ac.in (U. B. Jani), ardeshanabhavik@gmail.com (B. A. Ardeshana), ajaympatel2003@yahoo.com (A. M. Patel), anandyjoshi@gmail.com (A. Y. Joshi)

ABSTRACT

In the current manuscript, an approach to modeling and simulation of single-wall carbon nanocones (SWCNC) has been suggested for mass sensing applications. Finite element modeling and dynamic analysis of SWCNC with cantilever beam boundary condition, various disclination angles of 60°, 120°, 180°, 240°, 300° and 10, 15, 20 A° lengths have been completed using atomistic molecular structure. This study has been conducted to evaluate and identify the difference in fundamental frequencies shown by these nanodevices when subjected to sensing applications. The study also displays the outcome of alteration in the length of nanocones on the vibrational frequencies. It is witnessed that increasing length of a SWCNC with the same apex angle outcomes in a drop in the fundamental frequency. Additionally, it is clear from the outcomes that SWCNC with greater apex angles displays lesser values of fundamental frequencies. Original and defective single-walled nanocones have been analyzed to study the effect of defects like vacancy defect and Stone-Wales defect. The results show the fact that with the change in the disclination angle and defects there is a significant amount of variation in the stiffness due to the different position on defects in nanocones. The outcomes propose that smaller lengths of nanocones are good contenders for sensing applications as they display extensive variation in the fundamental frequencies. It also shows that as the mass increases a certain limit.

4.1 INTRODUCTION

In current periods, divergent nanostructures such as carbon nanotubes (CNTs) [1], carbon nanocones (CNCs) [4, 5], fullerenes, [2], and carbon nanorings [3] have generated an excessive amount of interest for calculating the properties. By reason of varied, possible uses of single-wall carbon nanocones (SWCNCs) in diverse regions, for example, cold electron and field emitter [6], adsorbent [7], and mechanical sensors [8, 9], a complete understanding of their mechanical, physical, and electronic properties are needed.

The cones with various summit edges of 19.2°, 38.9°, 60, 84.6°, and 112.9° were tentatively created by Krishnan et al. [10]. Therefore, morphologies of the CNCs with dissimilar apex angles were studied by Naess et al. [11] using transmission electron microscopy (TEM), synchrotron X-ray, and electron diffraction.

Molecular dynamics (MD), virtual examination [12, 13], and density functional theory (DFT) [14] are the most extreme significant methodologies, while different methodologies can be segmented to Bernoulli-Euler/Timoshenko beam models [15–17], the shell models [18–20], the molecular structural models [21–24], and meshless methodologies [25, 26].

Yan et al. [27] contemplated the physical elements and adaptable properties of SWCNCs by utilizing the higher request continuum rule. They utilized every one of the five kinds of CNCs to test the impact of the conical angle on the mechanical properties. Lee and Lee had played out the model examination of SWCNTs and SWCNCs [28] utilizing the finite element method (FEM).

Wei et al. [29] observed the elastic and plastic properties of CNCs. They established that CNTs and CNCs bend correspondingly to the load till the bond-breaking starts and the results indicated that Young's modulus of the CNCs varies from 0.29 TPa to 0.73 TPa. Liao et al. [30] observed the tensile and compressive properties of CNCs. They simulated the buckling of the cones under compression by using the same MD scheme. They studied the influence of temperature, cone height, and apex angle on the tensile and compressive behavior of CNCs. Tsai and Fang [31] found the buckling properties of CNCs by using MD simulations. They calculated thermal stability and nanomechanics of closed-tip of CNCs by adopting MD simulation and the results indicated that the decreasing apex angle will increase total energy.

Yan et al. [32] calculated the mass sensitivity of SWCNCs with cantilevered boundary conditions. They proposed that the mass sensitivity of SWCNCs be a subject to the height and top radii of SWCNCs. Hu et al. [33] found the fundamental frequencies of SWCNC and the outcome indicated the sensitivity in length with the frequencies.

To detect the vibration characteristics, Ansari et al. [34] used FE-modeling results which were compared with MD simulations. To determine the resonant frequencies of CNTs, Yun and Park [35] used FEM and developed the Surface Cauchy-Born (SCB) technique for nanoscale material. Ardeshana et al. [36] used the nanocones for mass sensing applications. It has been detected that increasing the side length of a SWCNC with a constant apex angle results in a decrease in the fundamental frequencies. Ardeshana et al. [37] deal with the examination of double-walled carbon nanocones (DWCNCs) for resonance-based applications. They have done a dynamic analysis of DWCNCs with cantilever boundary conditions and different disclination angles through atomistic molecular structure modeling. Results show that with the alteration in the disclination angles and defects there was a considerable amount of difference in the stiffness owing to the change in the bond orientation of the nanocones.

In the current script, authors have studied the vibrational characteristics of SWCNC. The authors have also examined the vibrational characteristics of SWCNCs for the mass-sensing application. Two diverse kinds of faults, i.e., vacancy defect and Stone-Wales defect have been incorporated in the model of SWCNCs. Here authors have fabricated cantilever boundary conditions alongside the above-stated faults to confirm the usage of SWCNC for mass sensing applications.

4.2 GEOMETRIES CONSTRUCTION OF SWCNCs

Naturally, cones have a big range of apex angles as reported by Jaszczak et al. [38], which explains the disclination cone helix structure. The cone-helix model is effective in forecasting the anticipated apex angles of graphite cones produced under several laboratory settings and made naturally from fluids during metamorphism shape. In relation to the condition of the Eulers theorem and graphene sheet construction, there are five apex angles that are geometrically acceptable as stated by Lin et al. [39].

Nanocones possess a pointy tip that can be useful to find out the specific mechanical properties of nanocones. Nanocones are very useful material for different types of technological applications. The angle of the sector removed from a flat graphene sheet to form a cone is known as disclination angle. Nanocones are classified according to their disclination angle. In this study, the authors have analyzed carbon nanocone of four different disclination angle 60°, 120°, 180° and 240° with three different lengths of cones are 10 Å, 15 Å, and 20 Å. Cone sheets with disclination angles of 60°, 120°, 180°, and 240° are shown in Figure 4.1.

The graphene sheet was cut in a different shape to create a circular graphene layer. This graphene layer was then revolved by touching the angular gap, that's how the nanocone is created. In the same way, other graphene angles are cut with different angles to create nanocones with different disclamation angles.

The formation of different graphene sheets is shown in Figure 4.2. The apex angles of cones are 112.9°, 83.6°, 60°, and 38.9°, respectively for the disclination angle 60°, 120°, 180°, and 240°. The apex angle of cones is obtained as: $2\sin^{-1}(1-d_\theta/360)$.

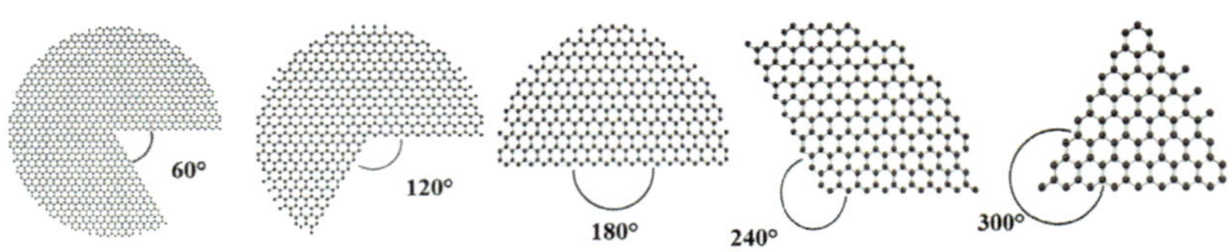

FIGURE 4.1 Graphene sheets with disclination angles of 60°, 120°, 160°, 240°, and 300° [36].

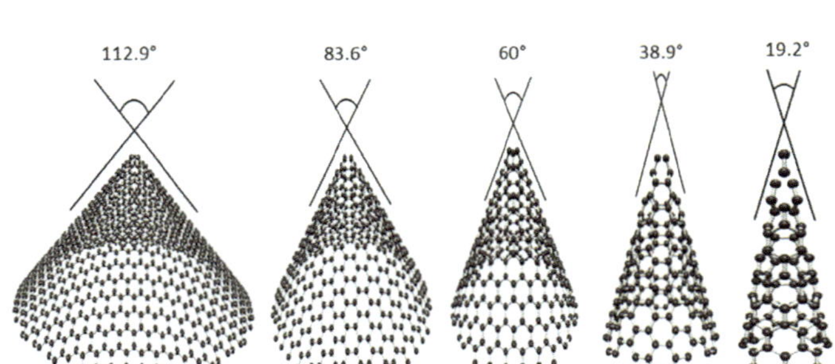

FIGURE 4.2 Cones with apex angles of 112.9°, 83.6°, 60°, 38.9°, and 19.2° [36].

Polar coordinates of 2D graphene sheets are then converted into Cartesian coordinates of 3D nanocone. Conversion of graphene into a cone is shown in Figure 4.3 with a disclamation angle of 240°. The polar coordinates of point P, indicated in graphene division as (l, α) and the parameters of the graphene sector are L and φ. Sector OAB is bent at apex O where $\overline{OA}$ and $\overline{OB}$ lengths are the same. So, A point of $\overline{OA}$ length is touch to point B of $\overline{OB}$ length [32].

Considering the point P(X, Y, Z), the following equations are obtained [40]:

$$X = r\cos\beta,\ Y = r\sin\beta,\ Z = -\sqrt{l^2 - r^2} \tag{1}$$

where, β and r are the unknown parameters. The graphene sheet transforms from the three-dimensional cone and for that the relation between the corresponding angles as below:

$$\frac{\beta}{2\pi} = \frac{\alpha}{\phi} \tag{2}$$

From Figure 4.3 the angle between $O'^2O'C'$ and $O'^2O'A'$ is the dihedral angle β, which can be found by rearranging the above Eq. (2) as below:

$$\beta = \alpha\left(\frac{2\pi}{\varphi}\right) \tag{3}$$

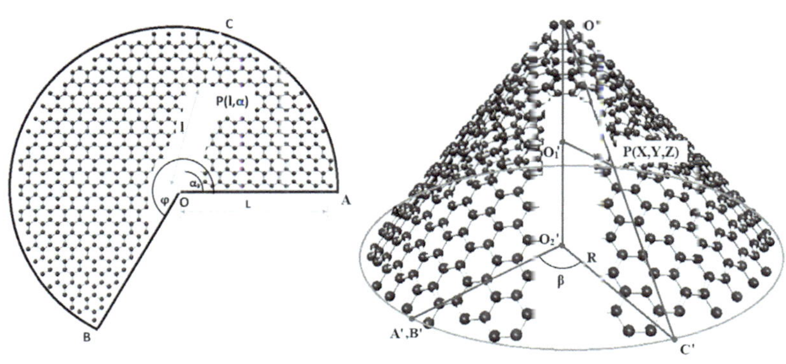

FIGURE 4.3 Symbols and variables used in the transformation from a graphene sheet for a cone with a disclination angle 240° [36].

The relation between radiuses r and R with lengths l and L are as follow:

$$\frac{r}{R}=\frac{l}{L} \tag{4}$$

For obtaining the base radius R of cone by below Eq. (5)

$$R=\frac{L\varphi}{2\pi} \tag{5}$$

For finding the radius r, substitute the value of R from Eq. (5) to Eq. (4)

$$r=\frac{l}{2\pi} \tag{6}$$

For determining the point P in the three-dimensional cone just by substitute the value r and β from Eqns. (6) and (3), respectively in Eq. (1) as follow:

$$X=\left(\frac{l\varphi}{2\pi}\right)\cos\left[\alpha\left(\frac{2\pi}{\varphi}\right)\right],\ Y=\left(\frac{l\varphi}{2\pi}\right)\sin\left[\alpha\left(\frac{2\pi}{\varphi}\right)\right],\ Z=-l\sqrt{1-\left(\frac{\varphi}{2\pi}\right)} \tag{7}$$

Here, X, Y, and Z indicate the atom coordinate of nanocones and angles are in radians.

4.3 MOLECULAR STRUCTURAL MECHANICS (MSM) BASED MODELING OF SWCNC

In the MSM approach, covalent bonds in between atoms are considered as beam elements. The stiffness of this beam elements have been considered from applying the process of potential energy. The total force on independently atomic nuclei is the summation of the force formed by the electrons and electrostatics force between the absolutely charged nuclei themselves. The overall formulation for the potential energy is

$$\Pi=\sum U_r+\sum U_\theta+\sum U_\phi+\sum U_\omega \tag{8}$$

where, U_r is the energy because of bond stretch interaction, U_θ the energy because of bending (bond angle variation), U_Φ the energy because of dihedral angle torsion, $U\omega$ the energy because of out-of-plane torsion.

$$U_r = \frac{1}{2}K_r(r - r_0)^2 = \frac{1}{2}K_r(\Delta r)^2 \tag{9}$$

$$U_\theta = \frac{1}{2}K_\theta(\theta - \theta_0)^2 = \frac{1}{2}K_\theta(\Delta\theta)^2 \tag{10}$$

$$U_\tau = U_\phi + U_\omega = \frac{1}{2}K_\tau(\Delta\phi)^2 \tag{11}$$

where, K_r, K_θ, and K^τ are the bond stretching, bond bending, and torsional resistance force constants, correspondingly, while Δr, $\Delta\theta$, and $\Delta\phi$ represent bond stretching increment, bond angle variation, and angle variation of bond twisting, respectively.

As the potential energy in the two approaches is independent, energy equivalence of the stored energy of the two approaches, that is, molecular mechanics and structural mechanics [37]

$$\frac{EA}{L} = K_r, \frac{EI}{L} = K_\theta, \frac{GJ}{L} = K_\tau \tag{12}$$

The elastic properties of the beam element are given as [37]

$$D = 4\sqrt{\frac{k_\theta}{k_r}}, \quad E = \frac{k_r^2 L}{4\pi k_\theta}, G = \frac{k_r^2 k_\phi L}{8\pi k_\theta^2} \tag{13}$$

where, D, L, E, I, and G represent the diameter, length of cone, Young's modulus, the moment of inertia, and shear modulus of the beam element.

4.4 MODAL ANALYSIS OF SWCNC

In this study, the matrix equations are taken as the dynamic three-dimensional lumped model system. The equation of motion in a generalized form can be written as:

$$[M]\{\ddot{v}\} + [B]\{\dot{v}\} + [K]\{v\} = [F] \tag{14}$$

where, [M] indicate the global mass matrix, $\{\ddot{v}\}$ indicates the second time derivative of the displacement vector $\{v\}$ (i.e., the acceleration), $\{\dot{v}\}$ indicates the velocity vector, and [B], [K], [F] indicates global damping matrix, global stiffness matrix, and force vector, respectively.

Modal analysis can be used to determine mode shape and natural frequency. Damping effects have been overlooked in the current study. So, the new equation of motion for the undamped system is as given:

$$[M]\{\ddot{v}\}+[K]\{v\}=\{0\} \tag{14a}$$

Any pre-stress is not included in the structural stiffness matrix [K]. Free vibration is harmonic for the linear system which can be considered in the following form:

$$\{u\}=\{\phi\}_i \cos \omega_i t \tag{15}$$

where, vector $\{\phi\}_i$ represents the mode shape of ith natural frequency, the vector ω_i in radians per unit time represents the ith natural circular frequency and t represents the time.

Rewriting Eq. (14a) we get:

$$\{-\omega_i^2[M]+[K]\}\{\phi\}_i=\{0\} \tag{16}$$

There are two possibilities which satisfy the equality of the above equation, the first choice is $\{\phi\}_i$ = {0} and the second choice is ([K] $-\omega^2$ [M]) = 0. The first choice is insignificant and, therefore, our attention is not on it. The second choice gives the following answer:

$$\left|[K]-\omega^2[M]\right|=0 \tag{17}$$

This is an eigenvalue problem that may be solved for up to n values of ω^2 and n eigenvectors $\{\phi\}_i$ which satisfy Eq. (15), where n is the number of DOFs. The eigenvalue and eigenvector extraction techniques are used in the Block Lanczos method. Rather than outputting the natural circular frequencies $\{\omega\}$, the natural frequencies (f) are output as

$$f_i=\omega_i/2\pi \tag{18}$$

where, f_i is the ith natural frequency (cycles per unit time). Normalization of each eigenvector $\{u\}_i$ to the mass matrix is performed according to

$$\{\phi\}_i^T [M]\{\phi\}_i = 0 \tag{19}$$

In the normalization, $\{\phi\}_i$ is normalized such that its largest component is 1.0 (unity). The natural frequency of a structure is related to its geometry, mass, and boundary conditions. For the nanocones considered here, the mass was assumed to be that of each carbon atom, 2.0×10^{-26} kg, and the rotational degrees of freedom of the atom are neglected.

In the finite element modeling, the BEAM 188 element in ANSYS was used to simulate the carbon bonds and the carbon atoms were simulated as the mass element of type MASS21. The authors have assumed that the cross-sections of the beam elements were uniform and circular, and the necessary input data of the BEAM 188 element were Young's modulus E, the Poisson's ratio m, and the diameter of the circular cross-section d taken from Table 4.1.

TABLE 4.1 The Input Data Prepared for the BEAM188 and MASS21 Elements [36]

Division	Current Work
Beam Type	Euler-Bernoulli
C-C bond length	0.142 nm
Poisson's ratio	0.3
Density	2.3×10^{-21} g/nm^3
Young's modulus	5448 nN/nm^2
Mass of carbon atom	1.994×10^{-23} g

4.5 RESULT DISCUSSION

In this study, researchers have a model of SWCNCs with disclination angle 60°, 120°, 180°, 240°, and 300°, and lengths of 10Å, 15Å, and 20Å. The mass has been put on the top of the cone whose values vary from 10^{-21} to 10^{-16} to check the variation caused by the mass on the natural frequency and mode shape of the SWCNC. The variation in natural frequency versus the value of the mass added for SWCNC for all disclination angle and 10Å length are shown in Figure 4.4. The same graph for all disclination angles and 15Å, 20Å length is given in Figures 4.5 and 4.6. As observed from the graphite seems that the value of frequency does not change for a large amount of mass but as mass increases up to 10^{-20} a significant increase in the frequency is found. From this graphite can be seen that SWCNC can be used for a very small amount of mass sensing.

The deformed shape of nanocones with 20 Å length and the four different disclination angles 60° are illustrated in Figure 4.7. The variations in natural frequency versus the mode of vibration for the cantilever condition of nanocones with disclination angles 60°, 120°, 180°, 240° for all length of 20 Å is compared in Figures 4.4–4.7.

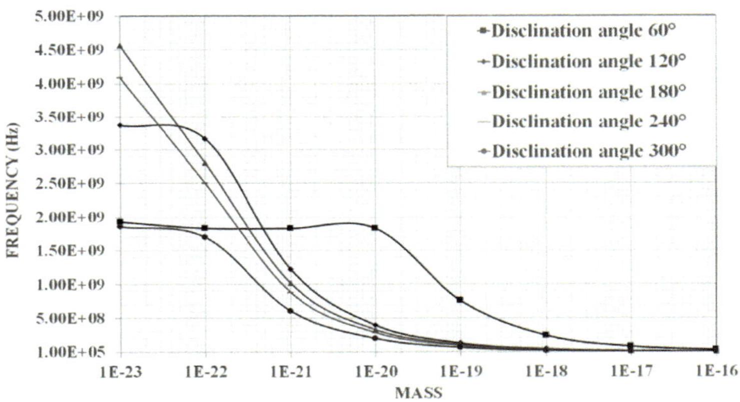

FIGURE 4.4 Mass sensing of SWCNCs with considering 10Å length for different disclination angles of 60°, 120°, 180°, 240°, and 300°.

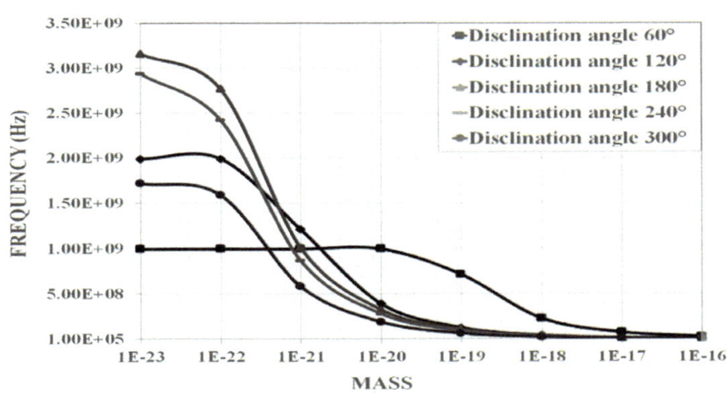

FIGURE 4.5 Mass sensing of SWCNCs with 15Å length cone for different disclination angles of 60°, 120°, 180°, 240°, and 300°.

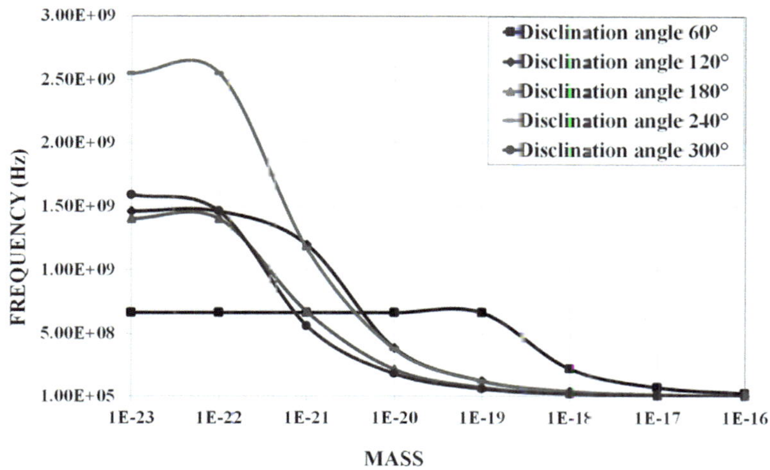

FIGURE 4.6 Mass sensing of SWCNCs with 20 Å length cone for different disclination angles of 60°, 120°, 180°, 240°, and 300°.

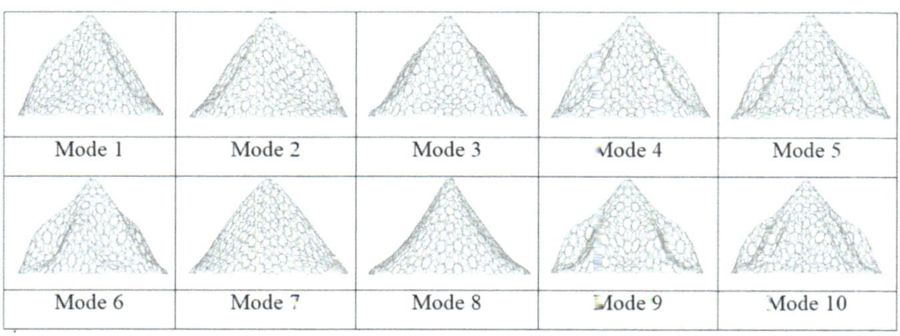

FIGURE 4.7 First 10 mode shapes of SWCNC with 10^{-20} mass attached to the top of the cone and disclination angle of 60°.

Change in the disclination angle does not make a significant change in the fundamental frequencies for higher mass. While a change in the disclination angle makes the variation of frequencies very high for smaller mass. As the graph of disclination angle, 60° shows a change in frequencies when mass is 10^{-19}. After 10^{-20} frequencies, variation is high at the mass of 10^{-21} to 10^{-23}.

From all the above-stated outcomes, it seems that pristine SWCNC can be used for mass sensing of a very small amount of mass as frequencies vary slightly at high mass value.

In all lengths of SWCNC for the same disclination angle value of frequencies for mass 10^{-16} to 10^{-19} remains almost equal. So these masses are hard to recognize in mass sensing. While small mass like 10^{-20} to 10^{-23} has higher frequencies variation with the length and disclination angle. So that can be easily recognized on the frequency scanner it shows in Figures 4.8–4.12.

Authors have evaluated frequencies for deferent types of defects and there location of the defect on the surface of SWCNC as shown in Figure 4.12 such as vacancy defect and Stone-Wales defect. For producing the vacancy defect author removed one carbon atom at a particular location such as the top, middle, and bottom of the cone and associated three C-C bonds. For this defective SWCNC mass attached has been kept constant 10^{-16} to check only for defect variations. Defects recreated on length of 20Å and on four disclination angle including 60°, 120°, 180°, and 240° which are shown in Figures 4.12 and 4.13.

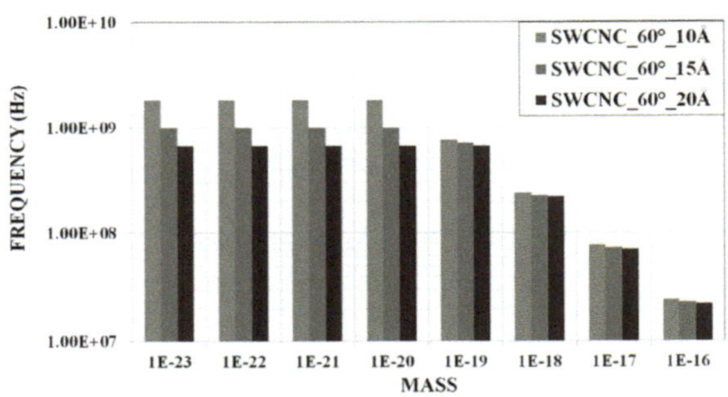

FIGURE 4.8 Mass sensing of SWCNCs with disclination angle 60° for different length of 10Å, 15Å, and 25 Å.

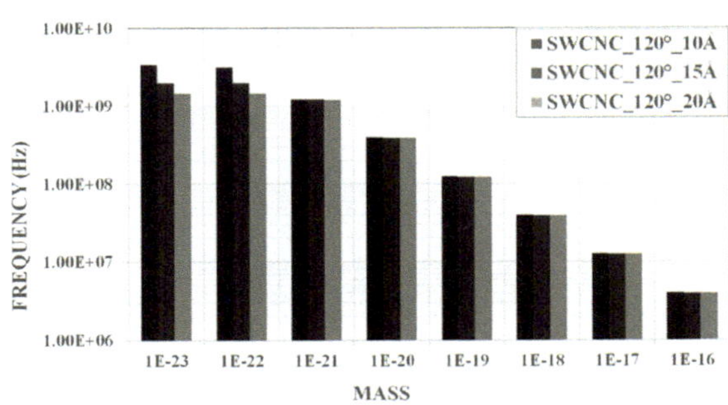

FIGURE 4.9 Mass sensing of SWCNCs with disclination angle 120° for different length of 10Å, 15Å, and 25 Å.

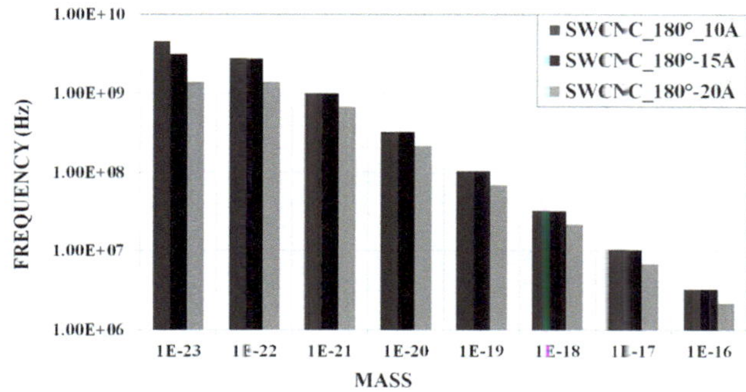

FIGURE 4.10 Mass sensing of SWCNCs with disclination angle 180° for different length of 10Å, 15Å, and 25 Å.

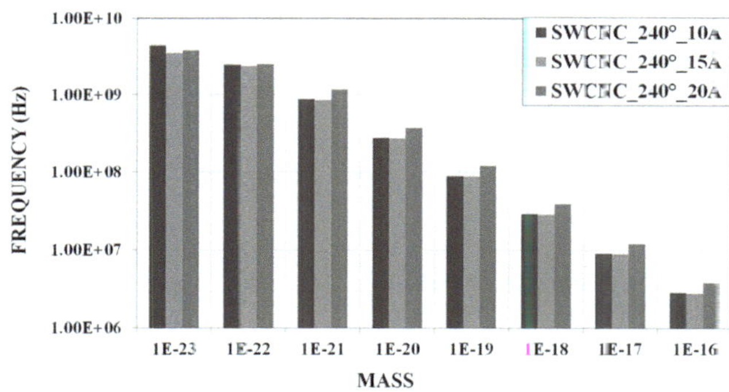

FIGURE 4.11 Mass sensing of SWCNCs with disclination angle 60° for different length of 10Å, 15Å, and 25 Å.

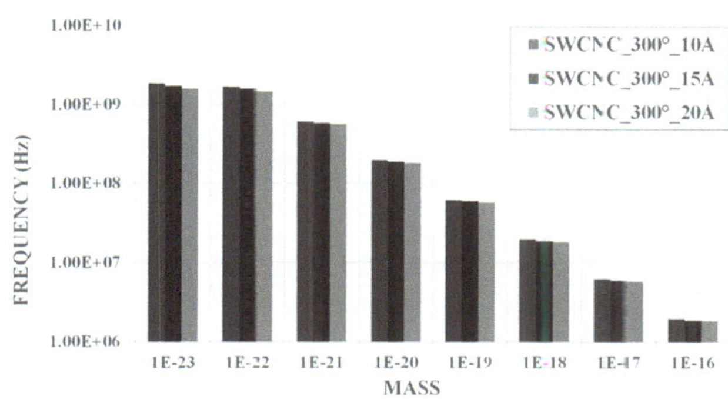

FIGURE 4.12 (a) Vacancy defect at top of SWCNC, (b) Vacancy defect at the middle of SWCNC, (c) Vacancy defect at bottom of SWCNC, (d) Vacancy defects at the top, middle, and bottom of SWCNC with disclination angle 240° and height 20 Å.

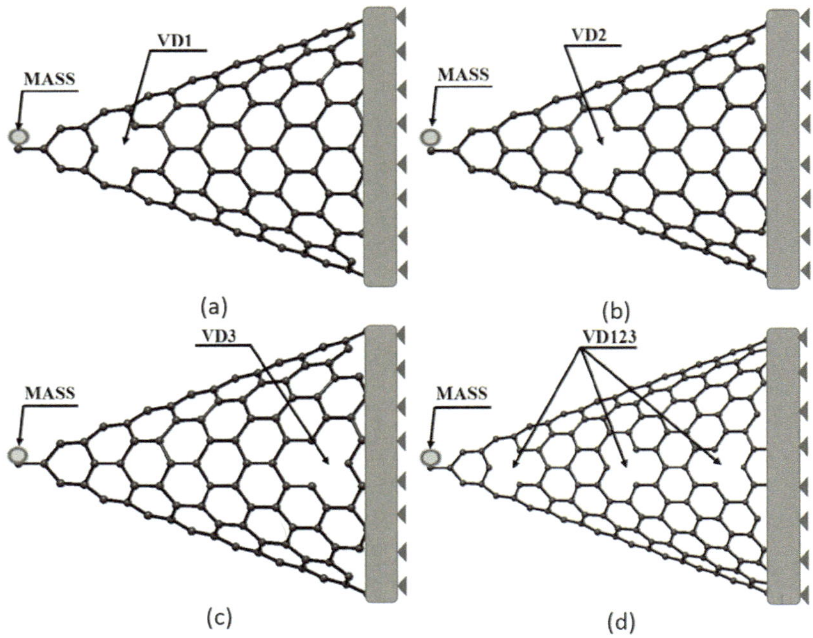

FIGURE 4.13 Schematic models of mass sensing with Stone-Wales defect in SWCNC with disclination angle 240° and height 20 Å.

The deformed shape of nanocones with 20 A° length and the four different disclination angles 120°.

Figures 4.14–4.17 demonstrates the mode shapes of SWCNC with a disclination angle 120° for the attached mass 10^{-16} gm and considering single and multiple vacancy defects. It is sensed that for greater modes, the localized bending effects of the apex to-base circle of the SWCNCs are increasing. It is witnessed that the difference in axial, twist, and bending mode frequencies in Figure 4.14. Figures 4.15–4.17 shows a similar kind of trends.

The graph of frequencies versus all kinds of vacancies for different disclination angles is shown in Figure 4.18. By comparing the graph of single vacancy defect at different locations, multiple vacancy defects and Stone-Wales defect it can be seen that frequencies increase with disclination angle.

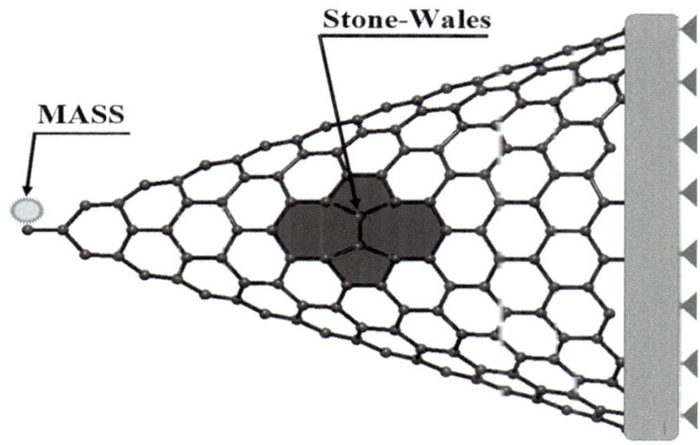

FIGURE 4.14 First 10 mode shapes of SWCNC with disclination angle 120° by considering vacancy defect 1 and 10^{-16} gm attached mass on the tip of the cone.

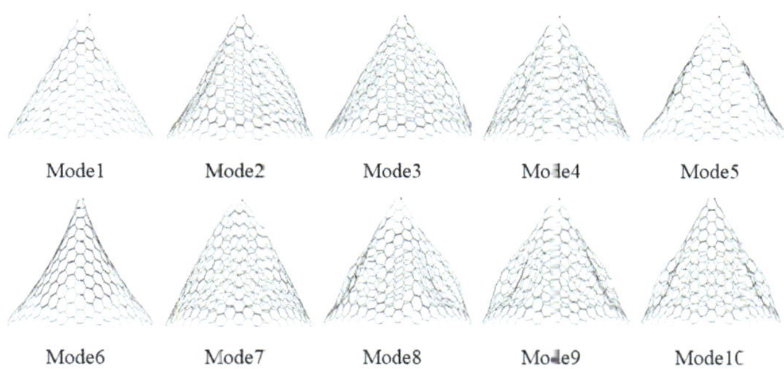

FIGURE 4.15 First 10 mode shapes of SWCNC with disclination angle 120° by considering vacancy defect 1 and 10^{-16} gm attached mass on the tip of the cone.

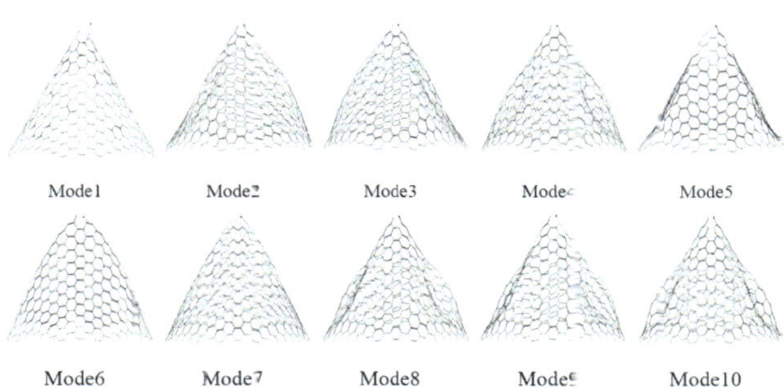

FIGURE 4.16 First 10 mode shapes of SWCNC with disclination angle 120° by considering vacancy defect 1 and 10^{-16} gm attached mass on the tip of the cone.

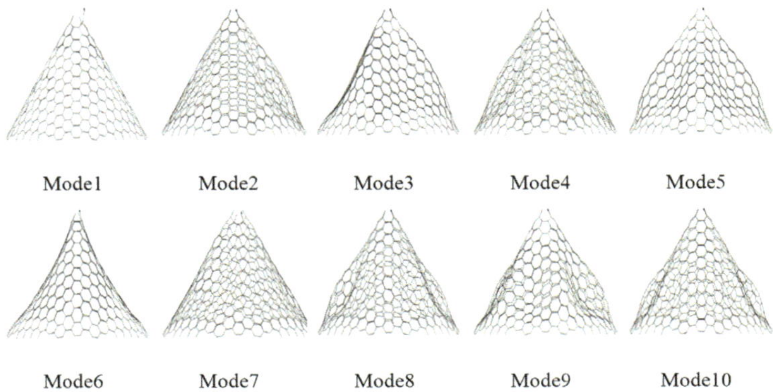

FIGURE 4.17 First 10 mode shapes of SWCNC with disclination angle 120° by considering vacancy defect at three different locations and 10^{-16} gm attached mass on the tip of the cone.

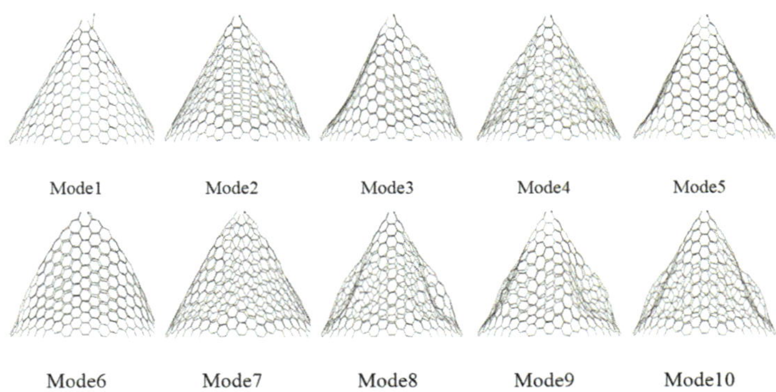

FIGURE 4.18 Frequency versus different types of defect for disclination angles 60°, 120°, 180°, and 240°.

4.6 CONCLUSIONS

1. Frequency does not change for a large amount of mass but as an increase, mass up to 10^{-20} a significant increase in the frequency is found.
2. Change in the disclination angle does not make a significant change in the fundamental frequencies for higher mass.

3. The difference of axial, twist and bending mode frequencies is observed for a disclination angle 120° and for the attached mass 10^{-16} gm.
4. Single vacancy defect at different locations, multiple vacancy defects, and Stone-Wales defect it can be seen that frequencies increase with disclination angle.

KEYWORDS

- **disclination**
- **finite element method**
- **mass sensing**
- **single-walled carbon nanocones**
- **Stone-Wales defects**
- **vacancy**

REFERENCES

1. Iijima, S., (1991). Helical microtubules of graphitic carbon. *Nature, 354*, 56–58.
2. Kroto, H. W., Heath, J. R., O'Brien, S. C., Curl, R. F., & Smalley, R. E. (1985). C60: Buckminsterfullerene. *Nature, 318*, 162–163.
3. Kong, X. Y., Ding, Y., Yang, R., & Wang, Z. L., (2004). Single-crystal nanorings formed by epitaxial self-coiling of polar nanobelts. *Science, 303*, 1348–1351.
4. Ijima, S., Ichihashi, T., & Ando, Y., (1992). Pentagons, heptagons, and negative curvature in graphite microtubule growth. *Nature, 356*, 776–778.
5. Iijima, S., & Ichihashi, T., (1993). Single-shell carbon nanotubes of 1-nm diameter. *Nature, 363*, 603–605.
6. Yu, S. S., & Zheng, W. T., (2010). Effect of N/B doping on the electronic and field emission properties for carbon nanotubes, carbon nanocones, and graphene nanoribbons. *Nanoscale, 2*, 1069–1082.
7. Majidi, R., & Ghafoori, T. K., (2010). Study of neon adsorption on carbon nanocones using molecular dynamics simulation. *Physica. B: Condensed Matter, 465*, 2144–2148.
8. Hu, Y., Liew, K. M., He, X. Q., Li, Z., & Han, J., (2012). Free transverse vibration of single-walled carbon nanocones. *Carbon, 50*, 4418–4423.
9. Yan, J. W., Liew, K. M., & He, L. H., (2013). Ultra-sensitive analysis of a cantilevered single-walled carbon nanocone-based mass detector. *Nanotechnology, 4*.

10. Krishnan, A., Dujardin, E., Treacy, M. M. J., Hugdahl, J., Lynum, S., & Ebbesen, T. W., (1997). Graphitic cones and the nucleation of curved carbon surfaces. *Nature, 388*, 451–454.
11. Naess, S. N., Elgsaeter, A., Helgesen, G., & Knudsen, K. D., (2009). Carbon nanocones: Wall structure and morphology. *Science and Technology of Advanced Materials*.
12. Iijima, S., Brabec, C., Maiti, A., & Bernholc, J., (1996). Structural flexibility of carbon nanotubes. *Journal of Chemical Physics*, 2089–2092.
13. Yakobson, B. I., Campbell, M. P., Brabec, C. J., & Bernholc, J., (1997). High strain rate fracture and C-chain unraveling in carbon nanotubes. *Computational Materials Science, 8*, 341–348.
14. Sanchez-Portal, D., Artacho, E., Soler, J. M., Rubio, A., & Ordejon, P., (1999). Ab initio structural, elastic, and vibrational properties of carbon nanotubes. *Physical Review B—Condensed Matter and Materials Physics, 59*, 12678–12688.
15. Wang, C. M., Tan, V. B. C., & Zhang, Y. Y., (2006). Timoshenko beam model for vibration analysis of multi-walled carbon nanotubes. *Journal of Sound and Vibration, 294*, 1060–1072.
16. Hsu, J. C., Chang, R. P., & Chang, W. J., (2008). Resonance frequency of chiral single-walled carbon nanotubes using the Timoshenko beam theory. *Physics Letters A, 372*, 2757–2759.
17. Zhang, Y. Y., Wang, C. M., & Tan, V. B. C., (2009). Assessment of Timoshenko beam models for vibrational behavior of single-walled carbon nanotubes using molecular dynamics. *Advances in Applied Mathematics and Mechanics, 1*, 89–106.
18. Ru, C. Q., (2000). Effective bending stiffness of carbon nanotubes. *Physical Review B: Condensed Matter and Materials Physics, 62*, 9973–9976.
19. Yakobson, B. I., Brabec, C. J., & Bernholc, J., (1996). Nanomechanics of carbon tubes: Instabilities beyond linear response. *Physical Review Letters, 76*, 2511–2514.
20. Ru, C. Q., (2000). Elastic buckling of single-walled carbon nanotube ropes under high pressure. *Physical Review B, 62*, 10405–10408.
21. Odegard, G. M., Gates, T. S., Nicholson, L. M., & Wise, K. E., (2002). Equivalent-continuum modeling of nanostructured materials. *Composites Science and Technology, 62*, 1869–1880.
22. Li, C., & Chou, T., (2003). A structural mechanics approach for the analysis of carbon nanotubes. *International Journal of Solids and Structures, 40*, 2487–2499.
23. Rouhi, S., & Ansari, R., (2012). Atomistic finite element model for axial buckling and vibration analysis of single-layered graphene sheets. *Physica. E, 44*, 764–772.
24. Ansari, R., & Rouhi, S., (2010). Atomistic finite element model for axial buckling of single-walled carbon nanotubes. *Physica. E: Low Dimensional Systems and Nanostructures, 43*, 58–69.
25. Liew, K. M., Lei, Z. X., Yu, J. L., & Zhang, L. W., (2014). Post buckling of carbon nanotube-reinforced functionally graded cylindrical panels under axial compression using a meshless approach. *Computer Methods in Applied Mechanics and Engineering, 268*, 1–17.
26. Zhang, L. W., Lei, Z. X., Liew, K. M., & Yu, J. L., (2014). Static and dynamic of carbon nanotube-reinforced functionally graded cylindrical panels. *Composite Structures, 111*, 205–212.

27. Yan, J. W., Li, D., Liew, K. M., & He, L. H., (2012). Predicting mechanical properties of single-walled carbon nanocones using a higher-order gradient continuum computational framework. *Composite Structures, 94*, 3271–3277.
28. Lee, J. H., & Lee, B. S., (2012). Modal analysis of carbon nanotubes and nanocones using FEM. *Computational Materials Science, 51*, 30–42.
29. Wei, J. X., Liew, K. M., & He, X. Q., (2007). Mechanical properties of carbon nanocones. *Appl. Phys. Lett., 91*(26), 261906.
30. Liao, M. L., Cheng, C. H., & Lin, Y. P., (2011). Tensile and compressive behaviors of open-tip carbon nanocones under axial strains. *J. Mater. Res., 26*(13), 1577–1584.
31. Tsai, P. C., & Fang, T. H., (2007). A molecular dynamics study of the nucleation, thermal stability, and nanomechanics of carbon nanocones. *Nanotechnology, 18*(10), 105702.
32. Yan, J. W., Liew, K. M., & He, L. H., (2013). Ultra-sensitive analysis of a cantilevered single-walled carbon nanocone-based mass detector. *Nanotechnology, 24*(12), 125703.
33. Hu, Y. G., Liew, K. M., He, X. Q., et al., (2012). Free transverse vibration of single-walled carbon nanocones. *Carbon, 50*(12), 4418–4423.
34. Ansari, R., Momen, A., Rouhi, S., et al., (2014). On the vibration of single-walled carbon nanocones: Molecular mechanics approach versus molecular dynamics simulations. *Shock Vibr., 2014*.
35. Yun, G., & Park, H. S., (2008). A finite element formulation for nanoscale resonant mass sensing using the surface Cauchy-born model. *Comput. Methods Appl. Mech. Eng., 197*(41), 3324–3336.
36. Ardeshana, B., Jani, U., Patel, A., et al., (2017). An approach to modeling and simulation of single-walled carbon nanocones for sensing applications. *AIMS Materials Science, 4*(4), 1010–1028.
37. Ardeshana, B., Jani, U., Patel, A., et al., (2018). Characterizing the vibration behavior of double-walled carbon nanocones for sensing applications. *Materials Technology, 33*(7), 451–466.
38. Jaszczaka, J. A., Robinson, G. W., Dimovski, S., et al., (2003). Naturally occurring graphite cones. *Carbon, 41*, 2085–2092.
39. Cheng-Te, L., Chi-Young, L., Hsin-Tien, C., et al., (2007). Graphene structure in carbon nanocones and nanodiscs. *Langmuir, 23*, 12806–12810.
40. Wong, L. H., Zhao, Y., Chen, G., et al., (2006). Grooving the carbon nanotube oscillators. *Appl. Phys. Lett., 88*(18), 183107.

CHAPTER 5

Mechanics and Material Removal Modeling and Design of Velocity Transformers in Ultrasonic Machining

V. DHINAKARAN,[1] JITENDRA KUMAR KATIYAR,[2] and T. JAGADEESHA[3]

[1]*Center for Applied Research, Chennai Institute of Technology, Chennai, Tamil Nadu, India*

[2]*SRM Institute of Science and Technology, Chennai, Tamil Nadu, India*

[3]*Department of Mechanical Engineering, NIT Calicut, India, E-mail: jagdishsg@nitc.ac.in*

ABSTRACT

Ultrasonic machining is a nontraditional process, in which abrasives contained in the slurry are driven against the work by a tool oscillating at low amplitude (25–100 microns) and high frequency (15–30 kHz). It is employed to machine hard and brittle materials (both electrically conductive and non-conductive material) having hardness usually greater than 40 HRC. The process was first developed in the 1950s and was originally used for finishing EDM surfaces. In this chapter, a detailed process and its process parameters are discussed. A brief summary of the equipments and tool configurations is presented. The materials removal model for both ductile and brittle materials are discussed in great detail. Various velocity transformers design aspects and criteria are discussed in this chapter with practical examples.

5.1 INTRODUCTION

Ultrasonic refers to waves of high frequency above the audible range of 20 kHz. The ultrasonic machining was proposed by L. Balamuth in 1945. It was developed for the finishing of electro spark machine parts. The USM process consists of a tool made of ductile and tough material. The tool oscillates with high frequency and continuous abrasive slurry is fed between tool and workpiece. The impact of the hard abrasive particles fractures the workpiece thus removing the small particles from the work surface.

Ultrasonic machining is different from the conventional grinding process. Table 5.1 gives a comparison between the two [1–19].

TABLE 5.1 Comparison between Conventional Grinding and USM

Parameters	Conventional Grinding	USM
Motion	Motion of the grinding wheel is tangential to the workpiece	Motion of the abrasive particle is normal to the workpiece
Basic process	Material removal is by pure shear deformation	material removal occurs by shear deformation, brittle fracture through impact (hammering), cavitation, and chemical reaction
Abrasive grits	abrasive grits are bonded to the wheel	Abrasives are supplied externally in the form of slurry
Tool motion	Abrasive wheel is rotated by an electrical motor	Tool is vibrated using a magnetostriction effect which produces ultrasonic waves of high frequency.

5.2 DESCRIPTION OF PROCESS

Ultrasonic machining is a mechanical type of nontraditional machining process (Figure 5.1). It is employed to machine hard and brittle materials (both electrically conductive and non-conductive material) having hardness usually greater than 40 HRC. The process was first developed in the 1950s and was originally used for finishing EDM surfaces.

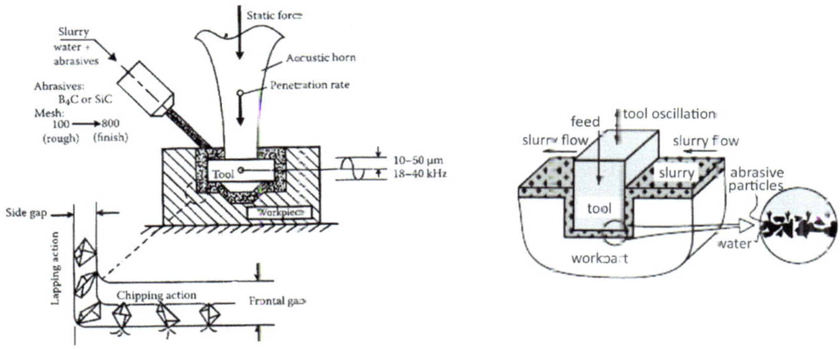

FIGURE 5.1 USM process.

In ultrasonic machining, a tool made of ductile and tough material and of the desired shape vibrates at ultrasonic frequency (19 to 25 kHz) with an amplitude of 15–50 microns over the workpiece. Generally, the tool is pressed down with a feed force F. Between the tool and work, the machining zone is flooded with hard abrasive particles generally in the form of a water-based slurry. As the tool vibrates over the workpiece, abrasive particles act as indenter and indent both work and tool material. Abrasive particles, as they indent, the work material would remove the material from both the tool and workpiece. In ultrasonic machining, material removal is due to crack initiation, propagation, and brittle fracture of the material. USM is used for machining hard and brittle materials, which are poor conductors of electricity and thus cannot be processed by electrochemical machining (ECM) or electro-discharge machining (EDM).

The tool in USM is made to vibrate with high frequency on to the work surface in the midst of the flowing slurry. The main reason for using ultrasonic frequency is to provide better performance. Audible frequencies of required intensities would be heard as extremely loud sound and would cause fatigue and even permanent damage to the auditory apparatus.

5.3 DESCRIPTION OF EQUIPMENT

The schematic diagram of the USM equipment is shown in Figure 5.2. The ultrasonic machining consists of:

1. High power sine wave generator;

2. Magnetostrictive transducer;
3. Tool holder;
4. Tool.

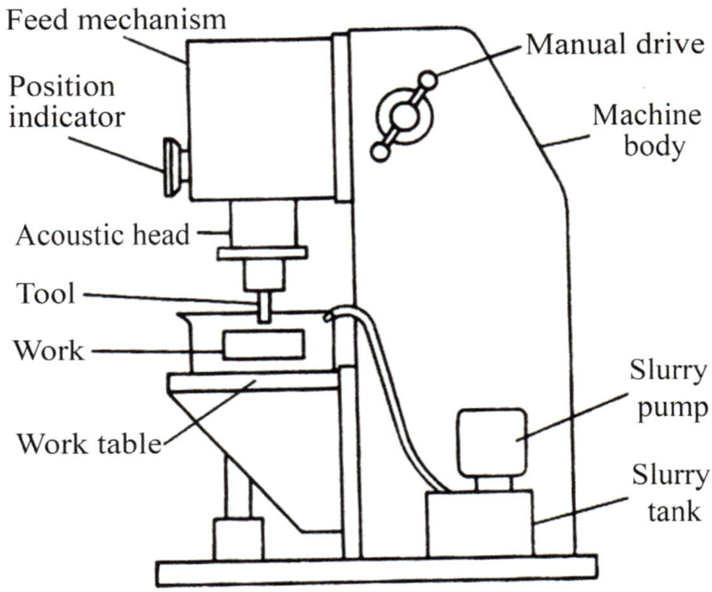

FIGURE 5.2 USM equipment.

5.3.1 HIGH POWER SINE WAVE GENERATOR

This unit converts low frequency (50/60 Hz) electrical power to high frequency (20 kHz) electrical power.

5.3.2 TRANSDUCER

Function of the transducer: The high-frequency electrical signal is transmitted to the transducer which converts it into a high frequency (15–20 kHz), low amplitude vibration (5 microns). The function of the transducer is to convert electrical energy to mechanical vibration using the principle of piezoelectric or magnetostriction.

There are two types of transducer used:

- Piezoelectric transducer; and
- Magnetostrictive transducer.

1. **Piezoelectric Transducer:** This transducer generates a small electric current when it is compressed, and also when the electric current is passed through the crystal it expands. When the current is removed, the crystal attains its original size and shape. Such transducers are available up to 900 Watts. Piezoelectric crystals have a high conversion efficiency of 95%.
2. **Magnetostrictive Transducer:** The magnetostriction effect was first discovered by Joule in 1874. According to this effect, in the presence of the applied magnetic field, ferromagnetic metals and alloys change in length. These transducers are made of nickel, nickel alloy sheets. Their conversion efficiency is about 20–30%. Such transducers are available up to 2000 Watts. The maximum change in length can be achieved is about 25 microns.

When the frequency of the ac signal provided by a high-frequency generator is tuned to the natural frequency of the transducer, resonance will occur. Because of the resonance amplitude of vibration increases. The transducer length is equal to half of the wavelength for the condition of resonance.

5.3.3 CONCENTRATORS (ACOUSTIC HORN)

1. **Function of the Concentrators (Figure 5.3):** The oscillation amplitude obtained from the magnetostrictive transducer is usually around 5 microns, which is too small for the removal of material from the workpiece. The function of the concentrator (also called mechanical amplifiers, Acoustic horn, and tool cone) is to amplify the amplitude of vibration of the magnetostrictive transducer from 5 microns to 40–50 microns. The concentrator also concentrates the power on a smaller machining area. To get the resonance condition, like transducer, the acoustic cone should be half-wavelength resonator.

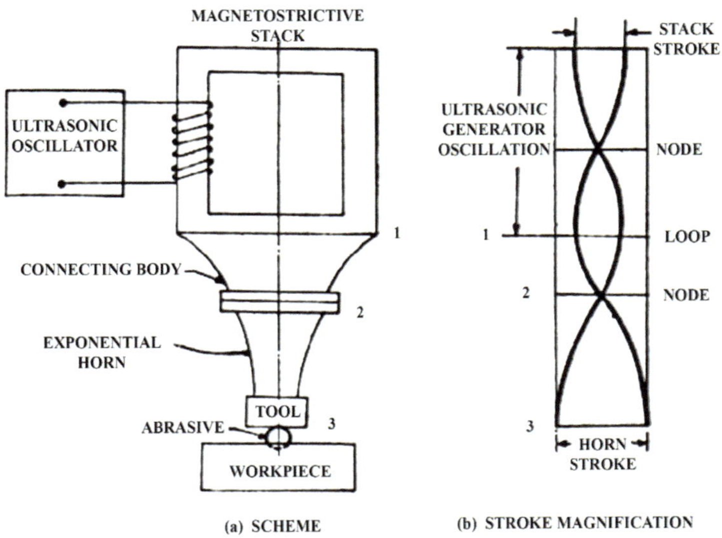

FIGURE 5.3 Concentrators used in USM.

The material of tools should have good acoustic properties, high resistance to fatigue cracking. Due measures should be taken to avoid ultrasonic welding between transducer and tool holder. Commonly used tool holders are monel, titanium, and stainless steel. Tool holders are more expensive, and demand higher operating costs.

The classification of the tool holder is given in Table 5.2.

TABLE 5.2 Classification of Tool Holders

Amplifying Tool Holder	Non-Amplifying Tool Holder
They give as much as 6 times increased tool motion. It is achieved by stretching and relaxing the tool holder material.	Non-amplifying tool holders have a circular cross-section and give the same amplitude at both ends.
MRR = 10 times the non-amplifying tool.	

5.3.4 TOOL

Tools are made of relatively ductile materials like brass, stainless steel, or mild steel so that the tool wear rate (TWR) can be minimized. The value of the ratio of TWR and MRR depends on the kind of abrasive, work material, and tool materials.

The design considerations for the tool are:

- The tool is made up of strong but ductile metal.
- Stainless steels and low carbon steels are used for making the tools.
- Aluminum and brass tools wear ten and five times faster than steel tools.
- The geometrical features are decided by the process.
- The diameter of the circle circumscribed about the tool should not be more than 1.5–2.0 times the diameter of the end of the concentrator.
- The tool should be as short and rigid as possible.
- When the tool is made hollow the internal contour should be parallel to the external one to ensure uniform wear.
- The thickness of any wall or projection should be at least five times the grain size of the abrasive.
- In the hollow tool, the wall should not be made thinner than 0.5–0.8 mm.
- When designing the tool consideration should be given to the side clearance which is normally of the order of 0.06–0.36 mm, depending on the grain size of the abrasive.

5.4 TOOL FEED MECHANISM

The feed mechanism of an ultrasonic machine must perform the following function:

1. Bring the tool very slowly to the workpiece to prevent breaking.
2. The tool must provide adequate cutting force and sustain cutting force during the machining operation.
3. The cutting force must be decreased when specified depth is reached.
4. Overrun a small distance to ensure the required hole size at the exit.

5. The tool has to come back to its initial position after machining is done.

There are four types of feed mechanism which are commonly used in USM:

- Gravity feed mechanism;
- Spring-loaded feed mechanism;
- Pneumatic or hydraulic feed mechanism; and
- Motor controlled feed mechanism.

1. **Gravity Feed Mechanism:** Figure 5.4 shows the operation of the gravity tool feed mechanism. In this mechanism counterbalance, weights are used to apply the required load to the head through pulley and rope arrangement. In order to reduce friction ball bearings are used. Gravity feed mechanisms are simple in construction but this mechanism is insensitive and inconvenient to adjust.

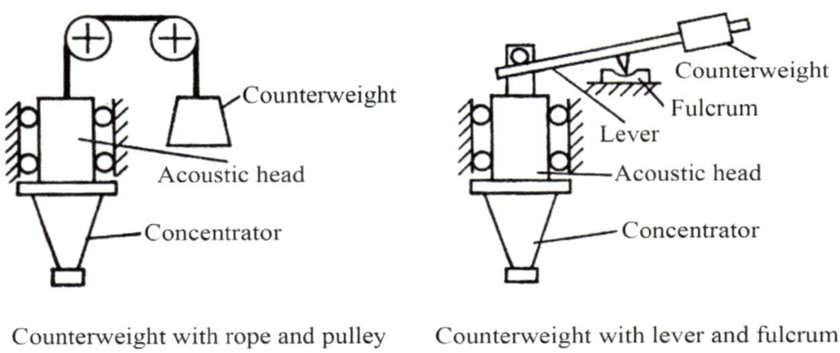

FIGURE 5.4 Gravity tool feed mechanism.

2. **Spring-Loaded Feed Mechanism:** Figure 5.5 shows the operation of the spring-loaded tool feed mechanism. In this mechanism, the spring pressure is used to feed the tool during the machining operation. This type of mechanism is quite sensitive and easy to adjust.

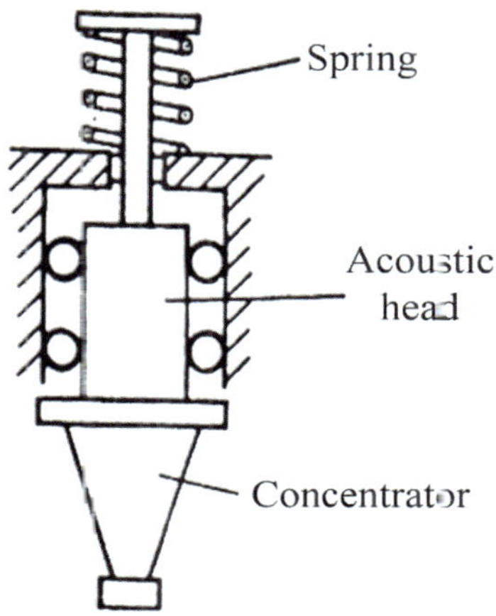

FIGURE 5.5 Spring-loaded tool feed mechanism.

3. **Pneumatic or Hydraulic Feed Mechanism:** Figure 5.6 shows the operation of pneumatic or hydraulic tool feed mechanism. In this mechanism, the hydraulic cylinder is used to give a linear motion of the tool. High feed rates and accurate positioning are possible with the hydraulic feed mechanism.

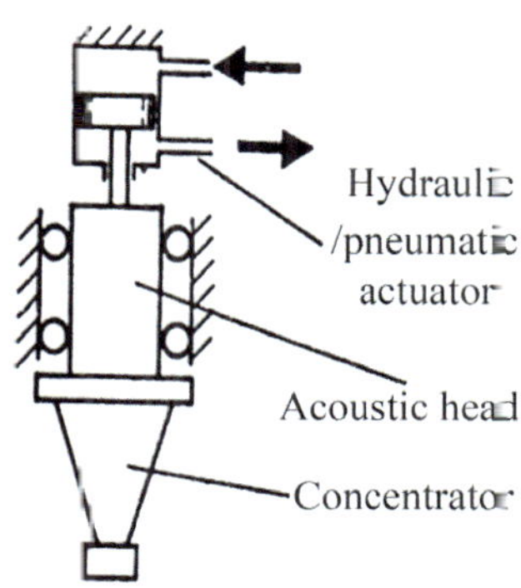

FIGURE 5.6 Pneumatic or hydraulic tool feed mechanism.

4. **Motor Controlled Feed Mechanism:** Figure 5.7 shows the operation of the motor controlled feed mechanism. This mechanism is used for precise control of the tool feed movement.

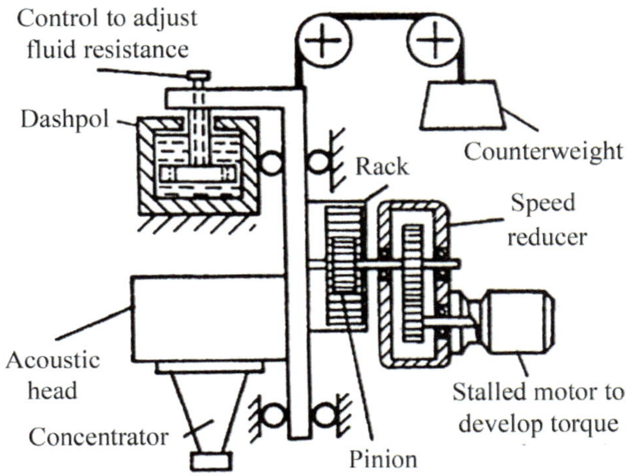

FIGURE 5.7 Motor controlled feed mechanism.

5.5 ABRASIVE SLURRY

In USM large variety of abrasive slurries are used. Some of them are:

- Boron carbide;
- Silicon carbide;
- Aluminum oxide; and
- Diamond dust.

Boron abrasive particles are used for machining tungsten, steel, and precious stones. Boon silica carbide is also used and it has 8–12% more abrasive than boron carbide. Alumina is used for machining ceramics, glass, and germanium. Alumina wears out very fast and loses its cutting power very fast. Silicon finds maximum application. Diamond and rubies are cut by diamond powder. Good surface finish, accuracy, and cutting rates is possible with diamond dust.

Proper selections of the abrasives are important. The selection of the abrasive particles depends on:

- Particle size;
- Hardness;
- Cost of abrasives;
- Durability of abrasives.

The life of abrasives depends on the hardness of abrasive material and work combinations. The higher life of abrasives can be obtained when the hardness of the abrasives are more than the hardness of the work material. The metal removal rate and surface finish depend on abrasive size particles. Coarse grains give higher MRR but the lower surface finish. Fine grains give good surface finish but the MRR is low.

The abrasive slutty is circulated by pumping and it requires cooling to remove the generated heat to prevent it from boiling in the gap and causing the undesirable cavitation effect. A refrigerated cooling system is provided to cool the slurry to a temperature of 5–6°C.

5.6 LIQUID MEDIA

In the USM process, the abrasive of about 30 to 60% by volume are suspended in a liquid medium. The several functions of the liquid medium are:

- Liquid medium acts as an acoustic bond between the vibrating tool and work.
- It carries the abrasive medium up to the cutting zone.
- It acts as a coolant and also carries waste abrasives and other swarf.
- It acts as transferring media for energy between the tool and workpiece.

The characteristics of the good suspension liquid medium are:

- The density of the liquid medium should be approximately equal to that of abrasive.
- The liquid medium should have good wetting characteristics. It should we tools, abrasives, and workpiece.
- The liquid medium should have high specific heat and thermal conductivity so that heat transfer between tool and work is effective.
- The liquid must have good flowability (low viscosity) and should carry the abrasives along with it.

- Liquid medium should not corrode workpiece, tool, and equipment.

Water is frequently used at the liquid carrier since it satisfies most of the requirements listed above. Some corrosion inhibitor is generally added to the water.

Ultrasonic vibrations imparted to fluid medium have the following important actions:

1. Ultrasonic vibrations will bring about the ultrasonic dispersions effect rapidly in the machining fluid medium between the tool end and the machining surface of the workpiece.
2. Ultrasonic vibrations bring violent circulations of the fluid as a result of ultrasonic micro agitation.
3. It causes the cavitation effect in the fluid medium arising out of the ultrasonic vibration of the tool in the fluid medium.

5.7 OPERATIONS OF ULTRASONIC MACHINING

The schematics of the USM operation is depicted in Figure 5.8. Ultrasonic machining is an economically viable process by which we can produce a cavity or hole in hard and brittle material. The sequence of operations is given below:

1. An electrical supply is given to a high-frequency generator. HF generator creates an alternating magnetic field which expands and contracts the stack made of magnetostrictive material (transducer). To get maximum magnetostriction, HF ac current is superimposed with dc pre-magnetizing current.
2. Since the frequency of the magnetic field created by the ac signal is the same as that of the natural frequency of the transducer, mechanical resonance occurs. The transducer length is equal to half of the length.
3. Oscillation of amplitude obtained from the transducer is about 5 microns, which is very small for metal removal. Therefore, it is amplified to 40–50 microns by fitting amplifiers into the output end of the transducer.
4. Acoustic horn transmits the mechanical energy to the tool and concentrates power on a small machining area.

5. Tool is fed into the workpiece by the automatic feed mechanism. It has a provision for measurement of static pressure exerted by the tool and penetration depth measurement.
6. Abrasive slurry under pressure is supplied to the working gap between tool and workpiece by a centrifugal pump.
7. Abrasive particles are hammered by the tool into the workpiece surface and they abrade the workpiece into the conjugate image of tool form.

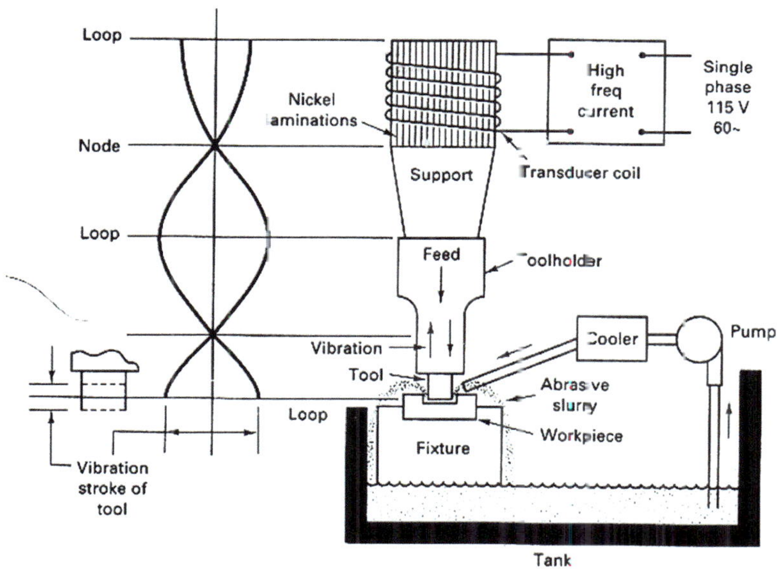

FIGURE 5.8 Operation of USM.

5.8 PROCESS PARAMETERS

1. Amplitude of vibration (15 to 50 microns);
2. Frequency of vibration (19 to 25 kHz);
3. Feed force (F) related to tool dimensions;
4. Feed pressure;
5. Abrasive size;
6. Abrasive material: Al_2O_3, SiC, B_4C, Boron silicarbide, Diamond;
7. Flow strength of the work material;
8. Flow strength of the tool material;

9. Contact area of the tool;
10. Volume concentration of abrasive in water slurry;
11. Tool:
    a. Material of tool;
    b. Shape;
    c. Amplitude of vibration;
    d. Frequency of vibration;
    e. Strength developed in tool;
    f. Gap between tool and work.
12. Work material:
    a. Material;
    b. Impact strength;
    c. Surface fatigue strength.
13. Slurry:
    a. Abrasive: hardness, size, shape, and quantity of abrasive flow;
    b. Liquid: Chemical property, viscosity, flow rate;
    c. Pressure;
    d. Density.

5.9 PROCESS CAPABILITY

1. Can Machine workpiece harder than 40 HRC to 60 HRC like carbides, ceramics, tungsten glass that cannot be machined by conventional methods. USM is not applicable to soft and ductile materials such as copper, lead, ductile steel, and plastics, which absorb energy by deformation.
2. Tolerance range 7 microns to 25 microns.
3. Holes up to 76 microns have been drilled hole depth up to 51 mm have been achieved easily. Hole depth of 152 mm deep is achieved by special flushing techniques.
4. Aspect ratio 40:1 has been achieved
5. Linear material removal rate –0.025 to 25 mm/min
6. Surface finish –0.25 micron to 0.75 micron
7. Non-directional surface texture is possible compared to conventional grinding
8. Radial over cut may be as low as 1.5 to 4 times the mean abrasive grain size.

5.10 APPLICATIONS

1. Machining of cavities in electrically non-conductive ceramics.
2. Used to machine fragile components in which otherwise the scrap rate is high.
3. Used for multistep processing for fabricating silicon nitride (Si_3N_4) turbine blades.
4. Large number of holes of small diameter. 930 holes with 0.32 mm have been reported using hypodermic needles.
5. Used for machining hard, brittle metallic alloys, semiconductors, glass, ceramics, sapphire, sintered carbides, etc.
6. Used for machining round, square, irregular shaped holes, and surface impressions.
7. Used in the machining of dies for wire drawing, punching, and blanking operations.
8. USM can perform operations like drilling, boring, sinking, blanking, grinding, trepanning, coining, engraving, and milling operations on all materials which can be treated suitably with abrasives.
9. USM has been used for the piercing of dies and for parting off and blanking operations.
10. USM enables a dentist to drill a hole of any shape on teeth without any pain.
11. Ferrites and steel parts, precision mineral stones can be machined using USM.
12. USM can be used to cut industrial diamonds.
13. USM is used for grinding quartz, glass, and ceramics.
14. Cutting holes with curved or spiral center lines and cutting threads in glass and mineral or metallo-ceramics.

5.11 ADVANTAGES OF USM

1. It can be used machine hard, brittle, fragile, and non-conductive material.
2. No heat is generated in work, therefore no significant changes in the physical structure of work material.

3. Non-metal (because of the poor electrical conductivity) that cannot be machined by EDM and ECM can very well be machined by USM.
4. It is burr less and distortion less processes.
5. It can be adopted in conjunction with other new technologies like EDM, ECG, ECM.
6. High accuracy with good surface finish can be achieved.
7. Posses the capability of drilling circular and non-circular holes in very hard materials like ceramics and other brittle materials.

5.12 DISADVANTAGES OF USM

1. Low metal removal rate.
2. It is difficult to drill deep holes, as slurry movement is restricted.
3. Frontal and side tool wear rate is high due to abrasive particles, especially when cutting steel and carbides. Side wear produces less accurate holes and cavities.
4. Tools made from brass, tungsten carbide, MS, or tool steel will wear from the action of abrasive grit with a ratio that ranges from 1:1 to 200:1.
5. USM can be used only when the hardness of work is more than 45 HRC.
6. It is not economical for soft material.
7. Not suitable for heavy stock removal.
8. USM is not useful for machining holes and cavities with a lateral extension of more than 25–30 mm with a limited depth of cut.
9. Every job needs a specific tool. Therefore, the tool cost is high.
10. The abrasive slurry should be changed regularly to replace worn-out particles. Therefore, the additional cost is involved.
11. Sharp corners are difficult to make using USM.

5.13 MECHANICS OF CUTTING IN USM

Theoretical analysis and experimental results have revealed that USM is a form of abrasion and material removal in the form of small grains by four mechanisms:

1. Throwing of abrasive grains.
2. Hammering of abrasive grains.
3. Cavitations in the fluid medium arising out of ultrasonic vibration of the tool.
4. Chemical erosion due to micro-agitation.

Material removal due to throwing and hammering is significant and MR due to cavitations and chemical erosion can be ignored.

Abrasive particles are assumed to be spherical in shape having diameter d_g. Abrasive particles move under high-frequency vibrating tools. There are two possibilities when the tool hits the particle.

- If the size of the particle is small and the gap between the tool and work is large, then the particle will be thrown by a tool to hit the workpiece; and
- If the size of the particle is large and the gap between tool and work is small, then the particle is hammered over the work surface.

The various mechanisms of the material removal process have been developed by a number of researchers. The important assumptions are:

1. All the abrasive particles are identical and spherical in shape.
2. All impacts are identical.
3. The metal removal rate is proportional to the volume of the work-piece per impact.
4. The metal removal rate is proportional to the number of particles making impact per cycle.
5. The rate of work material removal is proportional to the frequency. That is the number of cycles per unit time.

Consider an abrasive particle of diameter d_g hits the workpiece and generates an indentation of height δ as shown in the figure.

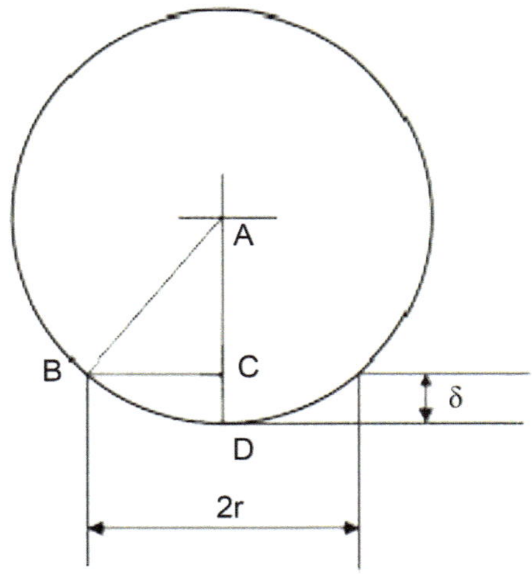

From the geometry,

$$AB^2 = AC^2 + BC^2$$

$$\frac{d_g}{2} = \left(\frac{d_g}{2} - \delta\right)^2 + r^2$$

$$\frac{d_g}{2} - \left(\frac{d_g}{2} - \delta\right)^2 = r^2$$

$$r^2 = \delta^2 + d_g\delta$$

Neglecting δ^2 term we can write:

$$r^2 = d_g\delta$$

$$r = \sqrt{d_g\delta}$$

Volume of the material removed is the volume of the hemispherical crater due to the fracture per grit per cycle.

$$\Gamma = \frac{1}{2}\left[\frac{4}{3}\pi r^3\right] = \frac{1}{2}\left[\frac{4}{3}\pi\left(r^2\right)^{\frac{3}{2}}\right] = \frac{1}{2}\left[\frac{4}{3}\pi\left(d_g\varepsilon\right)^{\frac{3}{2}}\right] = K_1\left(d_g\delta\right)^{\frac{3}{2}} \quad (1)$$

where, K_1 is constant.

Number of impacts (N) on the workpiece by the grits in each cycle depends on the number of grits beneath the tool at any time. This is inversely proportional to the diameter of grit.

$$N = \frac{K_2}{{d_g}^2} \quad (2)$$

where, K_2 is constant. All the abrasive particles under the tool need not be necessarily effective. MRR in AJM material can be expressed as:

$$\begin{aligned} MRR = &\{Volume\ of\ material\ removed\ per\ grit\ per\ cycle\} \\ &\times Number\ of\ impacts\ made\ by\ abrasives\ per\ second \\ &\times Number\ of\ impact\ per\ cycle \\ &\times Probability\ of\ abrasive\ particle\ under\ tool\ being\ effective \end{aligned}$$

$$MRR = \Gamma f \frac{K_2}{{d_g}^2} K_3$$

$$MRR = \left[K_1\left(d_g\delta\right)^{\frac{3}{2}}\right] f \frac{K_2}{{d_g}^2} K_3$$

$$MRR = \left[K_1 K_2 K_3 f \sqrt{\frac{(\delta)^3}{d_g}}\right] \quad (3)$$

Here, the Eq. (3) is the general MRR equation for USM.

5.13.1 MODEL 1: GRAIN THROWING MODEL

It is assumed that a particle is hit and thrown by the tool on to work surface. Assuming sinusoidal vibration, displacement of the tool (y) is given in the time period (t) and amplitude (a/2) of oscillation.

$$X = \frac{a}{2}\sin(2\pi ft)$$

By differentiating we get velocity V

Velocity = $V = \frac{a}{2}2\pi f \cos(2\pi ft)$

Maximum velocity $V_{max} = \pi af$

Let us assume that grits also move with the same velocity V_{max}, then we can write,

$$KE = \frac{1}{2}MV_{max}^{2} = \frac{1}{2}\left(\frac{\pi(d_g)^3}{6}\rho_g\right)(\pi af)^2 \quad (4)$$

Note: Here the diameter of the grit has to be taken to calculate the mass because we are here calculating the kinetic energy of the grit.

A grit penetrates to the depth equal to δ into the workpiece. The work done (WD) by the grit is given by:

$$\text{WD by the grit} = \frac{F\delta}{2} \quad (5)$$

Also, we know the flow strength of material = $\sigma_w = \frac{F}{\pi r^2}$

$$F = \sigma_w \times \pi r^2 \quad (6)$$

Using Eq. (6) in Eq. (5) we get:

$$\text{WD by the grit} = \frac{F\delta}{2} = \frac{\sigma_w \times \pi d_g \delta \times \delta}{2} \quad (7)$$

WD by the grit should be equal to the kinetic energy of the particle.

$$\frac{\sigma_w \times \pi d_g \delta \times \delta}{2} = \frac{1}{2}\left(\frac{\pi(d_g)^3}{6}\rho_g\right)(\pi af)^2$$

By simplifying, we have

$$\delta = \pi Afd_g\sqrt{\frac{\rho_g}{6\sigma_w}} \quad (8)$$

Using this Eq. (8) in the general equation (3), we have the volumetric material removal rate due to the throwing mechanism.

$$MRR = \left[K_1 K_2 K_3 d_g f^{\frac{5}{2}} \left[\frac{\rho_g (\pi a)^2}{6\sigma_w} \right]^{\frac{3}{4}} \right] \quad (9)$$

5.13.2. MODEL 2: GRAIN HAMMERING MODEL

When the gap between the tool and the workpiece is smaller than the diameter of the grit it will result in partial penetration in the tool (δ_t) as well as in the workpiece (δ_w). The values of (δ_t) and (δ_w) depends on the hardness of the tool and workpiece material, respectively. Force F acts on the abrasive particle only for a short time ΔT during the cycle time "T." During this time period, the abrasive particle is in contact with the tool and workpiece both. The mean force (F_{avg}) on the grit can be expressed by

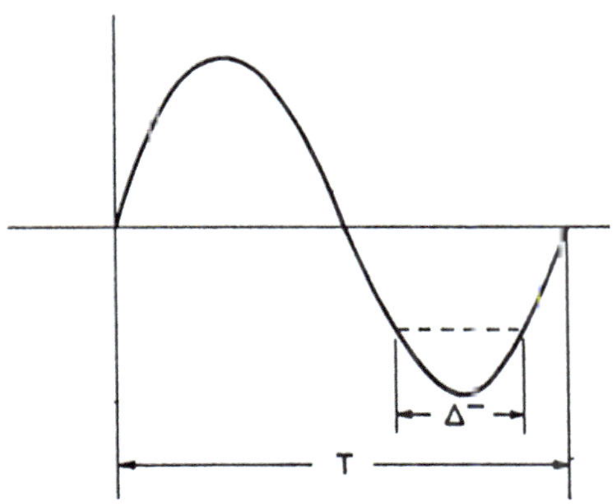

$$F_{avg} = \frac{1}{T} \int_0^T F(t)\, dt \quad (10)$$

Here F(t) is the force at any instant of time "t." Force on the grit by the tool starts increasing as soon as grit gets in contact with both the tool and the workpiece at the same time. It attains maximum value and then starts decreasing until attains the zero value. Hence, the momentum equation can be written as

$$\frac{1}{T}\int_0^T F(t)\,dt = \left[\frac{F}{2}\right]\Delta T \tag{11}$$

The position 'A' indicates the instant the tool face touches the abrasive grain (Figure 5.9). The period of movement from 'A' to 'B' represents the impact. The indentations, caused by the grain on the tool and the work surface at the extreme bottom position of the tool from the position 'A' to position 'B' are the total penetration.

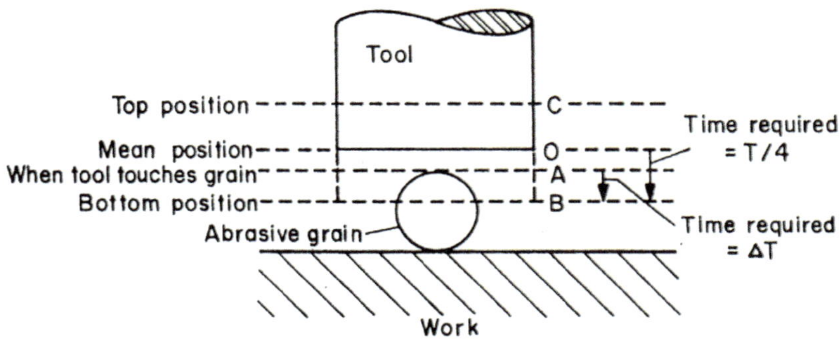

FIGURE 5.9 Grain hammering model.

Total penetration due the hammering is given by

$$\delta = \delta_w + \delta_t$$

where, a/2 is the amplitude of the oscillation of the tool. The mean velocity of the tool during the quarter cycle is given by $(a/2)/(T/4)$. Therefore, time (Δt) required to travel from A to B is given by the following equation:

$$\Delta T = \frac{\delta T}{2a} \tag{12}$$

Using Eq. (12) in Eq. (11)

$$F_{avg} = \frac{F\Delta T}{2T} = \frac{F\delta T}{2T \times 2a} = \frac{F\delta}{4a}$$

Or rewriting,

$$F = F_{avg}\left[\frac{4a}{\delta}\right] \tag{13}$$

Let N be the number of grains under the tool, stress acting on the tool σ_t, and the workpiece (σ_w) can be found as follows.

$$\sigma_w = \frac{F}{N\left(\pi d_g d_w\right)} \tag{14}$$

$$\sigma_t = \frac{F}{N\left(\pi d_g d_t\right)} \tag{15}$$

From Eqns. (14) and (15) we can write,

$$\frac{\sigma_w}{\sigma_t} = \frac{\delta_t}{\delta_w}$$

From the Eqns. (2), (13), and (14) we have

$$\sigma_w = \frac{F}{N\left(\pi d_g d_w\right)} = \frac{F_{avg} 4ad_g^2}{\delta\left(\pi d_g d_w\right)K_2} = \frac{F_{avg} 4ad_g}{K_2 \pi d_g \delta_w^{\ 2}\left[\frac{\delta_t}{\delta_w}+1\right]}$$

Writing $\frac{\delta_t}{\delta_w} = \lambda$ and rearranging we have

$$\delta_w = \sqrt{\frac{F_{avg} 4ad_g}{\sigma_w \pi K_2 (\lambda+1)}}$$

Volumetric material removal rate from the workpiece due to the hammering mechanism can be evaluated using the Eq. (3) as follows:

$$MRR = \left[K_1 K_2 K_3 f \sqrt{\left[\frac{(\delta_w)^3}{d_g}\right]}\right]$$

$$V_{hammering} = K_1 K_2 K_3 f d_g \left[\frac{F_{avg} 4a}{\sigma_w \pi K_2 (\lambda+1)}\right]^{\frac{3}{4}} \tag{16}$$

5.13.3 WORKED EXAMPLES

- **Problem 1:** Find out the approximate time required to machine a hole of diameter equal to 6.0 mm in a tungsten carbide plate (Flow strength of work material = 6.9×10^9 N/m^2) of thickness equal to one and half times of hole diameter. The mean abrasive grain size is 0.015 mm in diameter. The feed force is equal to 3.5 N. The amplitude of tool oscillations is 25 microns and the frequency is equal to 25 kHz. The tool material is copper having flow strength= 1.5×10^9 N/m^2 the slurry contains one part of the abrasives to one part of the water. Take the values of different constant as:

$K_1 = 0.3$, $K_2 = 1.8 \times 10^{-6}$(in SI units) and $K_3 = 0.6$ and abrasive slurry density = 3.8 g/cm^3. Also, calculate the ratio of the volume removed by the throwing mechanism to the volume removed by the hammering mechanism.

Data Given:

Diameter of the hole = 6 mm = 6×10^{-3} m

Depth of hole = 1.5 d = 9×10^{-3} m

Mean abrasive size (d_g) = 1.5×10^{-5} m

Feed force (F) = 3.5 N

Amplitude of oscillation = a/2 = 25×10^{-6} m

Frequency of oscillation = f = 25,000 CPS

Flow strength of work material = $\sigma_w = 6.9 \times 10^9$ N/m^2

Flow strength of tool material = $\sigma_t = 1.5 \times 10^9$ N/m^2

Abrasive grain density $\rho_g = 3.8 \times 10^3$ kg/m^3

$$\lambda = \frac{\textit{Flow Strength of work material}}{\textit{Flow Strength of tool material}} = \frac{\sigma_w}{\sigma_t} = 4.6$$

$K_1 = 0.3$, $K_2 = 1.8 \times 10^{-6}$ (In SI units) and $K_3 = 0.6$.

Solution:

1. **Grain Throwing Model:** Let us use equations that we have developed for the grain throwing model.

Penetration in workpiece due to throwing is given by:

$$\delta = \pi A f d_g \sqrt{\frac{\rho_g}{6\sigma_w}}$$

$$\delta = \pi\left(50\times10^{-6}\right)\left(2.5\times10^{4}\right)\left(1.5\times10^{-5}\right)\sqrt{\frac{3.8\times10^{3}}{6\times\left(6.9\times10^{9}\right)}} = 1.78\times10^{-5}\,mm$$

Volume removed by throwing is given by:

$$MRR = \left[K_1 K_2 K_3 f \sqrt{\left[\frac{(\delta_w)^3}{d_g}\right]}\right]$$

Substituting all the values we have:

$$MRR = \left[(0.3)\left(1.8\times10^{-6}\right)(0.6)\left(2.5\times10^{4}\right)\sqrt{\left[\frac{\left(1.78\times10^{-5}\right)^3}{1.5\times10^{-2}}\right]}\right] = 4.97\times10^{-3}\,mm^3/s$$

2. **Grain Hammering Model:** Penetration in workpiece due to hammering is given by:

$$\delta_w = \sqrt{\frac{F_{avg}\,4ad_g}{\sigma_w \pi K_2 (\lambda+1)}}$$

$$\delta_w = \sqrt{\frac{4\times3.5\times\left(2\times215\times10^{-6}\right)\left(1.5\times10^{-5}\right)}{\left(6.9\times10^{9}\right)\pi\left(1.8\times10^{-6}\right)(4.6+1)}}$$

$$\delta_w = 2.192\times10^{-4}\,mm$$

Volume removed by throwing is given by:

$$MRR = \left[K_1 K_2 K_3 f \sqrt{\left[\frac{(\delta_w)^3}{d_g}\right]}\right]$$

$$MRR = \left[(0.3)(1.8\times10^{-6})(0.6)(2.5\times10^{4})\sqrt{\left[\frac{(2.192\times10^{-4})^{3}}{(1.5\times10^{-2})}\right]} \right]$$

$$MRR = 0.2146\ mm^3/s$$

Total MRR = MRR due to throwing action + MRR due to hammering action

$$MRR = 4.97\times10^{-3}\frac{mm^3}{s} + 0.2146\frac{mm^3}{s} = 0.21957\frac{mm^3}{s}$$

Volume of the hole to be drilled = $\frac{\pi}{4}\times6^2\times9 = 254.416\ mm^3$

Time required to drill a hole = $\frac{Volume\ of\ the\ hole\ to\ be\ drilled}{Volumetric\ MRR\ due\ to\ both\ throwing\ and\ Hammering\ action}$

$$= \frac{245.416}{0.21957} = 1158.70\ seconds = 19.31\ Minutes$$

Ratio of the volumetric MRR due to throwing and hammering is given by:

$$\frac{MRR_{Throwing}}{MRR_{Hammering}} = \frac{0.00497}{0.2146} = 0.023$$

Thus, it is evident that the material removed by hammering is much more than by throwing (approximately 45 times) hence, for approximate calculations; MRR by throwing can be ignored.

- **Problem 2:** Determine the percentage change in the machining time for a USM operation cutting WC plates when the tool material is changed from copper to stainless steel. Take the ratio of flow stress of WC to flow stress of copper as 4.6 and the ratio of the flow stress of WC to flow stress of SST as 1.2.

Solution:

Let, the flow strength of work material = σ_w
Flow strength of tool material = σ_t

$$\lambda = \frac{Flow\ strength\ of\ work\ material}{Flow\ strength\ of\ tool\ material} = \frac{\sigma_w}{\sigma_t}$$

Penetration in workpiece due to hammering with copper as tool material is given by:

$$\delta_{w\text{-}copper} = \sqrt{\frac{F_{avg} 4ad_g}{\sigma_w \pi K_2 \left(\lambda_{copper} + 1\right)}}$$

Penetration in workpiece due to hammering with stainless steel (SST) as tool material is given by:

$$\delta_{w\text{-}SST} = \sqrt{\frac{F_{avg} 4ad_g}{\sigma_w \pi K_2 \left(\lambda_{SST} + 1\right)}}$$

Volume removed by throwing with copper as tool material is given by:

$$MRR_{copper} = \left[K_1 K_2 K_3 f \sqrt{\left[\frac{\left(\delta_{w-copper}\right)^3}{d_g}\right]}\right]$$

Volume removed by throwing with stainless steel as tool material is given by:

$$MRR_{SST} = \left[K_1 K_2 K_3 f \sqrt{\left[\frac{\left(\delta_{w-SST}\right)^3}{d_g}\right]}\right]$$

Assuming all other parameters do not change we can write:

$$\frac{MRR_{copper}}{MRR_{SST}} = \frac{\left[K_1 K_2 K_3 f \sqrt{\left[\frac{\left(\delta_{w-copper}\right)^3}{d_g}\right]}\right]}{\left[K_1 K_2 K_3 f \sqrt{\left[\frac{\left(\delta_{w-SST}\right)^3}{d_g}\right]}\right]}$$

$$\frac{MRR_{copper}}{MRR_{SST}} = \sqrt{\left[\frac{\left(\delta_{w-copper}\right)^3}{\left(\delta_{w-SST}\right)^3}\right]}$$

$$\frac{MRR_{copper}}{MRR_{SST}} = \sqrt{\left[\frac{\left(\sqrt{\frac{F_{avg}4ad_g}{\sigma_w \pi K_2\left(\lambda_{copper}+1\right)}}\right)^3}{\sqrt{\frac{F_{avg}4ad_g}{\sigma_w \pi K_2\left(\lambda_{SST}+1\right)}}^3}\right]} = \left\{\frac{(\lambda_{SST}+1)}{\left(\lambda_{copper}+1\right)}\right\}^{\frac{3}{4}}$$

$$\frac{MRR_{copper}}{MRR_{SST}} = \left\{\frac{(\lambda_{SST}+1)}{\left(\lambda_{copper}+1\right)}\right\}^{\frac{3}{4}} = \left\{\frac{\left(\frac{\sigma_w}{\sigma_{sst}}+1\right)}{\left(\frac{\sigma_w}{\sigma_{copper}}+1\right)}\right\}^{\frac{3}{4}} = \left\{\frac{(1.2+1)}{(4.6+1)}\right\}^{\frac{3}{4}} = 0.49622$$

$$\text{Time required for machining} = \frac{\textit{Volume of the hole to be drilled}}{\textit{Volumetric MRR due to both throwing and Hammering action}}$$

which gives:

$$\frac{MRR_{copper}}{MRR_{SST}} = \frac{machingtime_{SST}}{\left(machingtime_{copper}\right)}$$

Therefore, the percentage change in cutting time when the tool is changed from copper to SST is given by:

$$Percentage\ change = 1 - \frac{maching\ time_{SST}}{\left(maching\ time_{copper}\right)} = 1 - 0.49622 = 0.5037 \cong 50\%$$

Therefore, we can reduce the cutting time to 50% by changing the tool from copper to SST.

- **Problem 3:** Glass is being machined at a MRR of 6 mm^3/min by Al_2O_3 abrasive grits having a grit diameter of 150 microns, if 100-micron grits are to be used. What would be the MRR?

Solution:

Volumetric material removal rate from the workpiece due to the hammering mechanism can be evaluated using the Eq. (3) as follows:

$$MRR = \left[K_1 K_2 K_3 f \sqrt{\left[\frac{(\delta_w)}{d_g} \right]} \right]$$

$$V_{hammering} = K_1 K_2 K_3 f d_g \left[\frac{F_{avg}\, a}{\sigma_w \pi K_2 (\lambda + 1)} \right]^{\frac{3}{4}}$$

Assume that all the parameters remain the same we can write:

$$V_{hammering} = constant_1 d_g$$

Material removal rate for 150-micron grit is given by:

$$V_{hammering-150micron} = constant_1 \times 150 micron \qquad (1)$$

Material removal rate for 100-micron grit is given by

$$V_{hammering-100micron} = constant_1 \times 100 micron \qquad (2)$$

Also, it is given $V_{hammering-150micron} = 6 mm^3 / min$

Dividing Eqns. (1) by (2) we get

$$\frac{V_{hammering-150micron}}{V_{hammering-100micron}} = \frac{150}{100}$$

$$\frac{6 mm^3 / min}{V_{hammering-100micron}} = \frac{150}{100}$$

$$V_{hammering-100micron} = 4 mm^3 / min$$

Thus by decreasing the grit size, MRR decreases but surface finish increases.

➢ **Problem 4:** Glass is being machined at a MRR of 6 mm^3/min by Al_2O_3 abrasive grits having a grit diameter of 150 microns. The frequency of operation is 20 kHz. If the frequency is increased to 25 kHz. What would be the MRR?

Volumetric material removal rate from the workpiece due to the hammering mechanism can be evaluated using the Eq. (3) as follows:

$$MRR = \left[K_1 K_2 K_3 f \sqrt{\left[\frac{(\delta_w)^3}{d_g} \right]} \right]$$

$$V_{hammering} = K_1 K_2 K_3 f d_g \left[\frac{F_{avg} 4a}{\sigma_w \pi K_2 (\lambda + 1)} \right]^{\frac{3}{4}}$$

Assume that all the parameters remain the same we can write:

$$V_{hammering} = constant_1 f$$

Material removal rate for 150-micron grit is given by:

$$V_{hammering-150micron-old} = constant_1 \times f_{20} \tag{1}$$

Material removal rate for 100-micron grit is given by:

$$V_{hammering-150micron-new} = constant_1 \times f_{25} \tag{2}$$

Also, it is given $V_{hammering-150micron} = 6mm^3 / min$

Dividing Eqns. (1) by (2) we get:

$$\frac{V_{hammering-150micron-old}}{V_{hammering-150micron-new}} = \frac{constant_1 \times f_{20}}{constant_1 \times f_{25}}$$

$$\frac{6mm^3 / min}{V_{hammering-150micron-new}} = \frac{20kHz}{25kHz}$$

$$V_{hammering-150micron-new} = 7.5mm^3 / min$$

Thus by increasing the frequency, MRR increases.

5.14 MACHINING CHARACTERISTICS OF USM

The following are the USM process criteria:

1. Material removal rate;
2. Geometrical accuracy;
3. Surface finish;
4. Out of roundness.

Metal removal rate is affected by:

1. Work material;
2. Amplitude;
2. Frequency of tool oscillations;
4. Static pressure;
5. Abrasive concentration in the slurry. (Mixing ratio).

The accuracy and surface finish depends on:

1. Work materials;
2. Tool material and tool design;
3. Oscillation amplitude and grain size of the abrasives;
4. Hole depth and machining time;
5. Cavitation effect.

The following important parameters which affect MRR and surface finish are discussed in subsequent sections:

- Amplitude of tool oscillations;
- Frequency of tool oscillations;
- Abrasive grain size;
- Static loading (feed force);
- Effect of slurry concentration of the abrasive;
- Hardness ratio of the tool and the workpiece.

1. **Effect of Amplitude on MRR:** Increase in the amplitude of vibration increases MRR. To maximize the amplitude of the vibration concentrator should operate at the resonance frequency. Under certain circumstances, this limits also the maximum size of abrasive to be used.

When the amplitude of the vibration increases, the MRR is expected to increase. The actual nature of the variation is shown in Figure 5.10. There is some discrepancy in the actual values again. This arises from the fact that we calculated the duration of penetration Δt by considering average velocity.

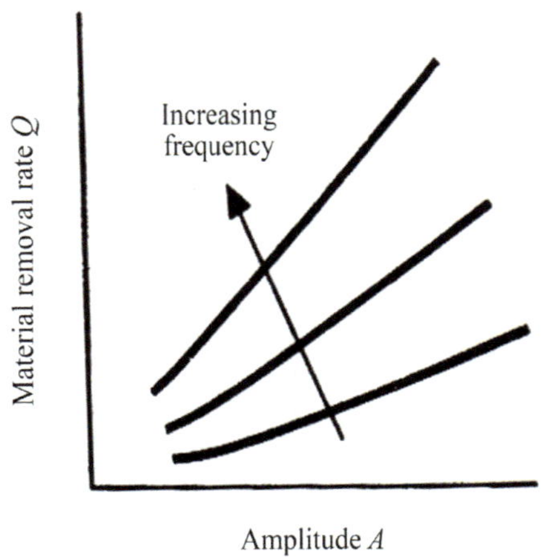

FIGURE 5.10 The actual nature of the variation.

Pressure also has an effect on the MRR. Figure 5.11 shows the effect of amplitude of vibration on MRR for different pressure.

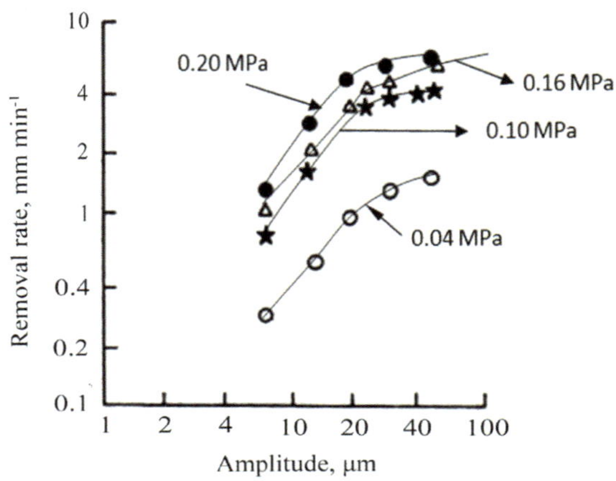

FIGURE 5.11 The effect of amplitude of vibration on MRR for different pressure.

2. **Effect of Frequency on MRR:** Frequency has a significant effect on MRR (Figure 5.12). Frequency used for the machining process must be resonant frequency to obtain the greatest amplitude at the tooltip and thus achieve the maximum utilization of the acoustic system. With an increase in the frequency of the tool head, the MRR should increase proportionally. However, there is a slight variation in the MRR with frequency

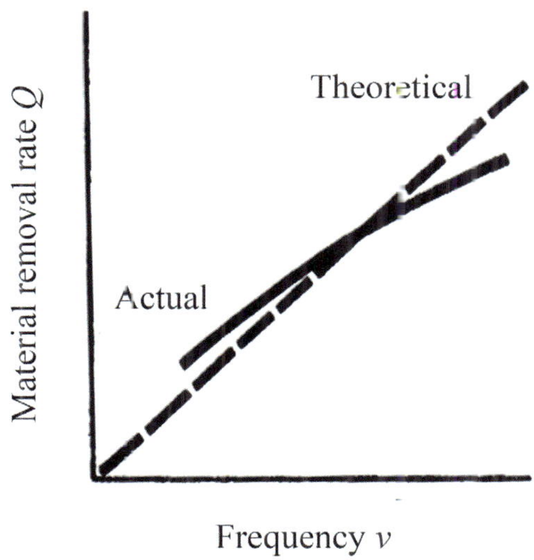

FIGURE 5.12 Effect of Frequency on MRR.

3. **Effect of Abrasive Grain Size:** MRR should also rise proportionately with the mean grain diameter. An increase in abrasive grain size results in higher MRR but poorer surface finish (Figure 5.13). Maximum MRR is achieved when the abrasive grain size is comparable with the amplitude of vibration of the tool. The hardness of the abrasives and method of introducing the slurry has also effect on MRR.

The concentration of the abrasives directly controls the number of grains producing impact per cycle. Figure 5.14 shows the effect of abrasive concentration on MRR. Silicon carbide gives lower MRR compared to boron carbide.

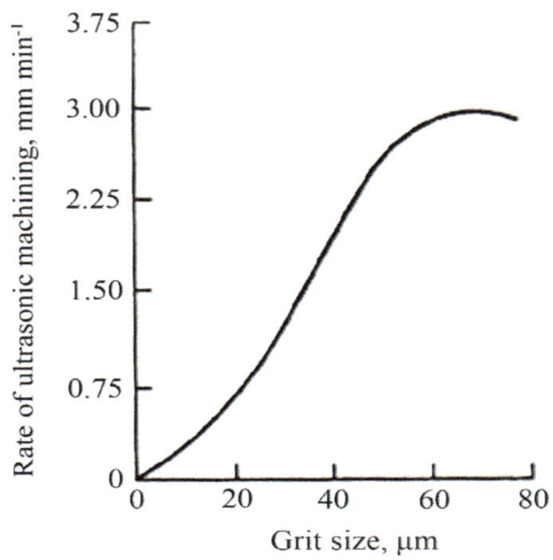

FIGURE 5.13 Effect of abrasive grain size.

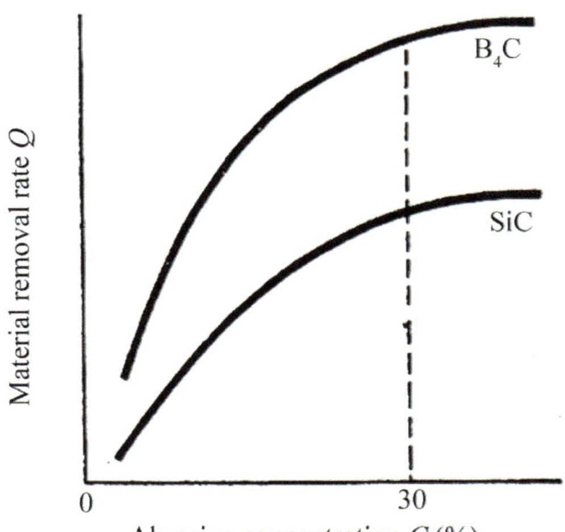

FIGURE 5.14 The effect of abrasive concentration on MRR.

Figure 5.15 shows that the surface finish is more sensitive to grain size in the case of glass which is softer than tungsten carbide. This is because in the case of a harder material the size of the fragments dislodged through a brittle fracture does not depend much on the size of the impacting particles.

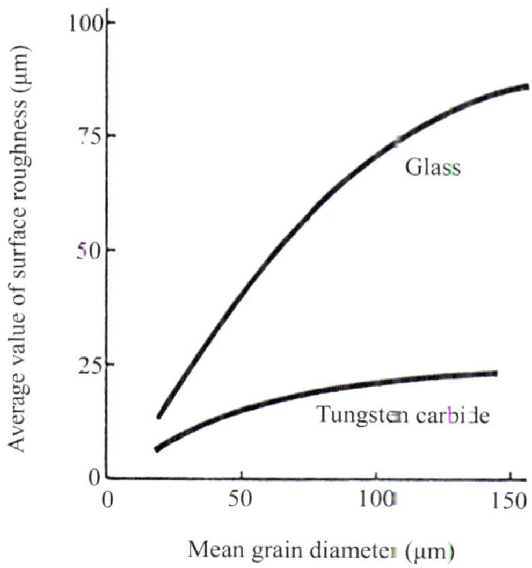

FIGURE 5.15 The surface finish is more sensitive to grain size in the case of glass.

4. **Effect of Applied Static Load (Feed Force):** MRR increase with the feed force (Figure 5.16). Maximum MRR depends on the amplitude of vibrations. The surface finish is found to be little affected by the applied static load. Higher loads, contrary to expectations, do not give a rougher finish. Surface finish, in fact, improves because the grains are crushed to small size with higher loads. However, a higher load gives lower MRR.

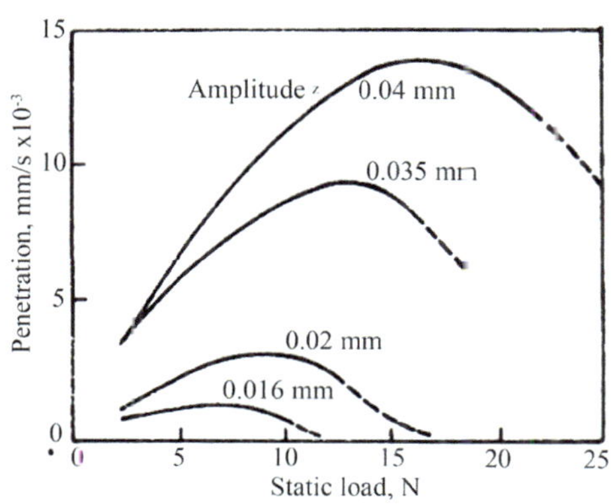

FIGURE 5.16 MRR increase with the feed force.

The variation of the metal removal rate for varying static load (feed force) is shown in Figure 5.17. As the tool size decreases, the penetration also increases.

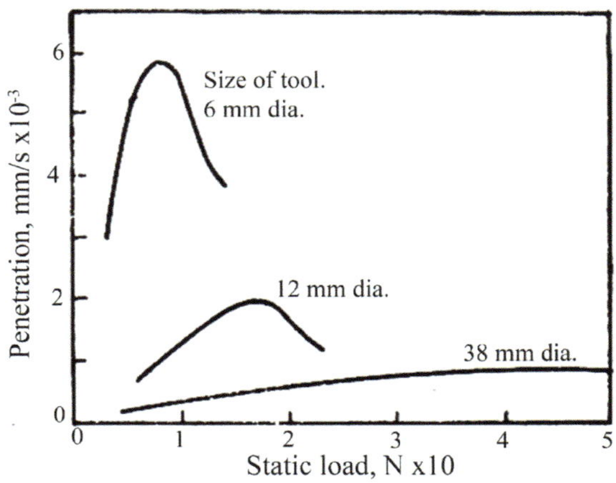

FIGURE 5.17 The variation of the metal removal rate for varying static load.

5. **Effect of Slurry, Tool, and Work Material:** MRR increases with slurry concentration. Slurry saturation occurs at 30 to 40% abrasive/water mixture (Figure 5.18).

The pressure with which the slurry is fed into the cutting zone affects MRR. In some cases, MRR can be increased even ten times by supplying the slurry at increased pressure.

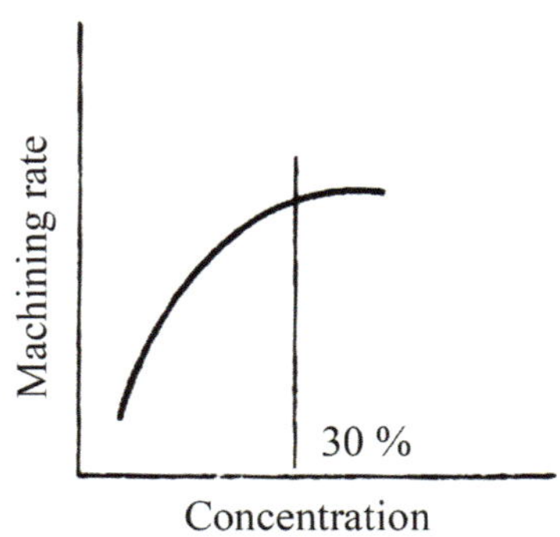

FIGURE 5.18 Effect of slurry, tool, and work material.

Apart from the process parameters some physical properties (e.g., viscosity) of the fluid used for the slurry also affects the MRR. Experiments show that MRR drops as viscosity increases (Figure 5.19).

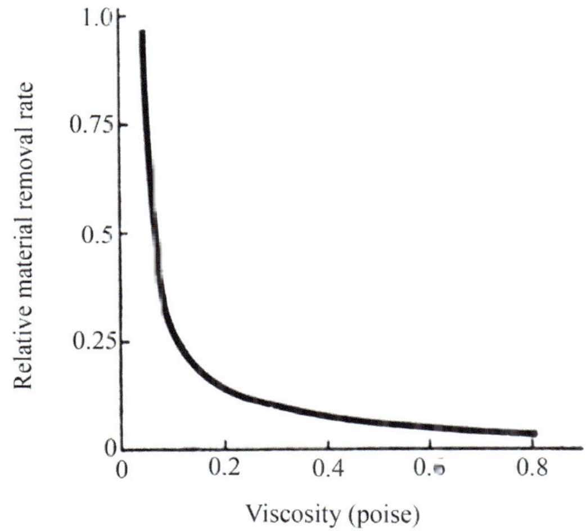

FIGURE 5.19 MRR drops as viscosity increases.

6. **Effect of Hardness Ratio of the Tool and the Workpiece:** The ratio of workpiece hardness and tool hardness affects the MRR quite significantly, and the characteristics are shown in Figure 5.20.

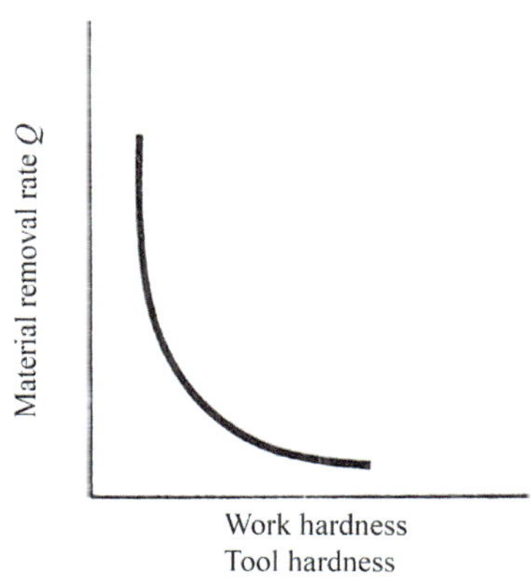

FIGURE 5.20 The ratio of work-piece hardness and tool hardness affects the MRR significantly.

The shape of the tool affects the MRR. Narrower rectangular tool gives more MRR compared to the square cross-section. The conical tool gives twice MRR compared to the cylindrical tool.

The brittle behavior of the material is important in determining the MRR. Brittle material can be cut at higher rates than ductile materials. Table 5.3 shows the relative MRR for different work materials. As can be seen, the more brittle material is machined more rapidly.

TABLE 5.3 Relative MRR for Different Material*

Work Material	Relative Removal Rate
Glass	100.00
Brass	6.6
Tungsten	4.8
Titanium	4.0
Steel	3.9
Chromium steel	1.4

*Table relative removal rate (Frequency of 16.3 kHz, grain diameter of 12.5 microns, and grain size of 100 mesh).

5.15 DESIGN OF HORN (VELOCITY TRANSFORMER)

The function of the horn (also called concentrator) is to amplify and focus the vibration of the transducer to an adequate intensity for driving the tool to fulfill the cutting operation. They are made of hard, non-magnetic, and easily machinable steel having good fatigue strength like K-Monel, metal bronze, and mild steel. Linearly tapered and exponentially taped horns have lengths equal to one-half of the wavelength of sound in the metal of which they are made.

A. **Design of Exponential Concentrator of Circular Cross-Section:** Let us consider the case of the exponential horn with a circular cross-section (Figure 5.21). This type of horn is used for cutting large diameter through holes in the workpiece.

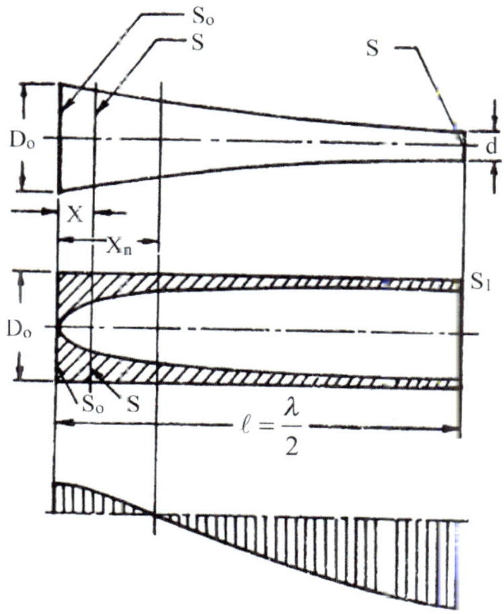

FIGURE 5.21 Exponential horn with a circular cross-section.

K = Transformation ratio – Amplification ratio:

$$K = \frac{D_o}{d} = \sqrt{\frac{S_o}{S_1}}$$

where, D_o is the larger diameter; d is the smaller diameter; S_o is the area of the larger section; and S_1 is the area of the smaller section.

Taper index for an exponential horn:

$$\beta = \frac{\ln(K)}{L}$$

The value of K should be chosen taking into consideration machining condition and the properties of magnetostrictive materials K = 3 to 4 for fine machining and 4 to 5 for rough machining.

The length of the horn or concentrator ***L*** is usually chosen as half-wave (n=1) or full-wave (n=2).

Length of the exponential horn is given by:

$$L = \frac{n\,C}{f}\sqrt{1+\left\{\frac{\ln(K)}{2\pi\,n}\right\}^2}$$

And the variation of D is given by:

$D = D_o \exp^{-\beta X}$ — this is also called law of change of shape.

A check must be made to ensure that operating frequency exceeds critical frequency f_c given by:

$$f_C = \frac{\beta C}{2\,\pi}$$

➢ **Problem 1:** Design a half wave steel exponentially tapered concentrator with a circular cross-section to work at a frequency of 20 kHz. The transducer has a larger diameter of 100 mm and a smaller diameter of 4 mm. Take the velocity of sound in steel C= 5×10^6 mm/sec.

Data Given

$D_O = 100$ mm
d = 4 mm
f = 20 kHz = 20,000 CPS.
$C = 5 \times 10^6$ mm/sec.
n = ½ (because it is a half-wave horn)
Schematic diagram is shown in Figure 5.22.

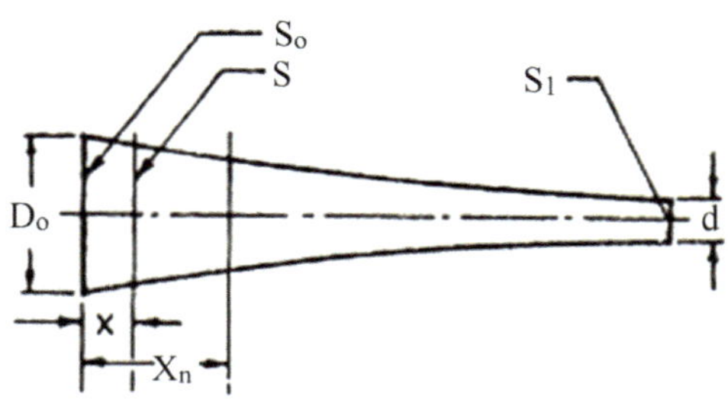

FIGURE 5.22 Schematic diagram for problem 1.

1. Concentrator ratio or amplification ratio or transformation ratio:

$$K = \frac{D_o}{d} = \frac{100}{4} = 20$$

2. **Length of the exponential horn:**

$$\frac{n\,C}{f}\sqrt{1+\left\{\frac{\ln(K)}{2\pi\, n}\right\}^2} = \frac{1}{2}x\frac{5x10^6}{20000}\sqrt{1+\left\{\frac{\ln(20)}{2\pi 12}\right\}^2} = 172.7\ mm$$

3. **Taper index for an exponential horn:**

$$\beta = \frac{\ln(K)}{L} = \frac{\ln(20)}{172.7} = 0.017344$$

4. **Law of change of shape is given by:**

$$D = D_o \exp^{-\beta X} = 100 \exp^{-0.017344}$$

5. **Check the critical frequency:**

$$f_C = \frac{\beta C}{2\,\pi} = \frac{0.017344}{2}x\frac{5x10^6}{\pi} = 13{,}803 Hz > \text{operating frequency of } 20{,}000\text{ Hz}$$

Design is safe; therefore, concentrator will work satisfactorily. The law of change of shape is given in Table 5.4.

TABLE 5.4 Change of Shape

X (mm)	0.017344X	$\exp^{(-0.017344X)}$	100 exp(–0.017344X)
0	0	1	100.00
10	–0.173	0.8408	84.08
30	–0.520	0.5943	59.43
50	–0.867	0.4201	42.01
70	–1.214	0.2970	29.70
90	–1.561	0.2099	20.99
110	–1.908	0.1484	14.84
130	–2.255	0.1049	10.49
150	–2.602	0.0742	7.42
170	–2.948	0.0524	5.24
173.7	–3.013	0.0492	4.92 (4 mm)

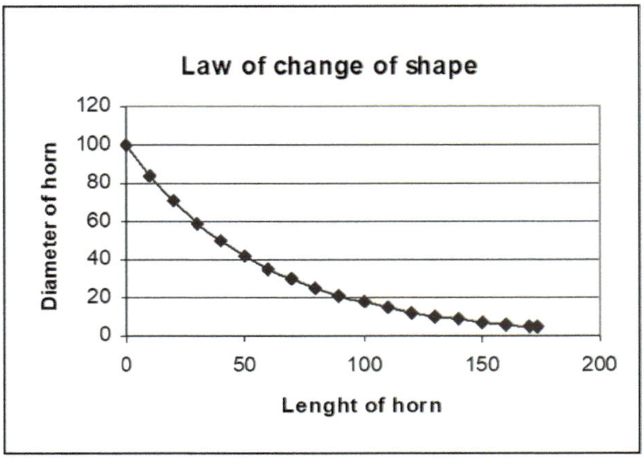

B. Design of Exponential Concentrator of Rectangular Cross-Section: Let us consider the case of the exponential horn with the rectangular cross-section. This type of horn is widely used in ultrasonic cutting (Figure 5.23).

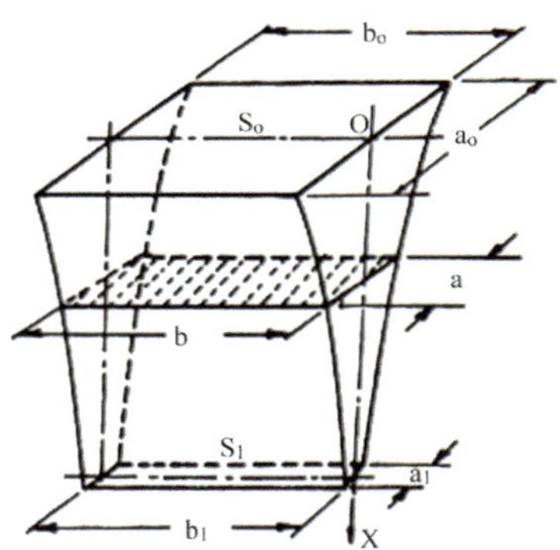

FIGURE 5.23 Exponential horn with the rectangular cross-section.
K= Transformation ratio – Amplification ratio

$$K = \sqrt{\frac{S_o}{S_1}} = \sqrt{\frac{a_o}{a_1}}$$

where, S_o is the area of the larger section; S_1 is the area of the smaller section; a_o and a_1 are widths at the larger and smaller end.

The taper index for an exponential horn is given by:

$$\beta = \frac{\ln(K)}{L}$$

The length of the horn or concentrator ***L*** is usually chosen as half-wave (*n*=1) or full-wave (*n*=2).

Length of the exponential horn is given by:

$$L = \frac{n\,C}{f}\sqrt{1+\left\{\frac{\ln(K)}{2\pi\,n}\right\}^2}$$

And the variation of D is given by:

$a = a_o \exp^{-2\beta X}$ – this is also called law of change of shape.

A check must be made to ensure that operating frequency exceeds critical frequency f_c given by:

$$f_C = \frac{\beta C}{2\,\pi}$$

- **Problem 2:** Design a HALF WAVE steel exponentially tapered concentrator with a rectangular cross-section to work at a frequency of 20 kHz. The transducer has a larger rectangular cross-section of 100 × 100 mm and a smaller rectangular section of 4 × 4 mm. Take the velocity of sound in steel C = 5 × 10^6 mm/sec.

Data Given

a_O = 100 mm

a_1 = 4 mm

f = 20 kHz = 20,000 CPS.

C = 5× 10^6 mm/sec.

n = ½ (because it is half-wave horn)

Schematic diagram is shown in Figure 5.24.

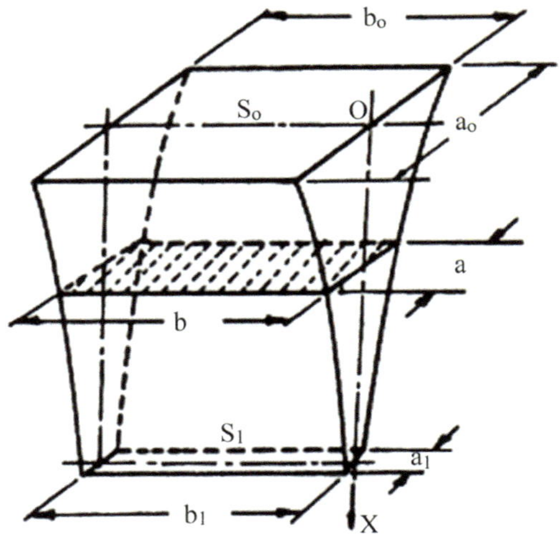

FIGURE 5.24 Schematic diagram for problem 2.

1. **Concentrator ratio or Amplification ratio or Transformation ratio:**

$$K=\sqrt{\frac{S_o}{S_1}}=\sqrt{\frac{a_o}{a_1}}=\sqrt{\frac{100}{4}}=4.47$$

2. **Length of the exponential horn:**

$$L=\frac{n\,C}{f}\sqrt{1+\left\{\frac{\ln(K)}{2\pi\,n}\right\}^2}\frac{1}{2}x\frac{5x10^6}{20000}\sqrt{1+\left\{\frac{\ln(4.47)}{2\pi 12}\right\}^2}=138.47\ mm$$

3. **Taper index for an exponential horn:**

$$\beta=\frac{\ln(K)}{L}=\frac{\ln(4.47)}{138.47}=0.0108138$$

4. **Law of change of shape is given by:**

$$a=a_o\ \exp^{-2\beta X}=100\exp^{-0.021627X}$$

5. **Check the critical frequency:**

$$f_C = \frac{\beta C}{2\pi} = 1.5X\frac{0.0108138}{2}x\frac{5x10^6}{\pi} = 1.5X8506 > \text{operating frequency}$$

of 20,000 Hz

Design is safe, therefore, concentrator will work satisfactorily.
Law of change of shape is given in Table 5.5.

TABLE 5.5 Change of Shape

X (mm)	0.021627X	exp $^{(-0.021627X)}$	100 exp $^{(-0.021627X)}$
0	0	1	100.00
10	–0.21627	0.8055	80.55
20	–0.43254	0.6489	64.89
40	–0.86508	0.4210	42.10
60	–1.29762	0.2732	27.32
80	–1.73016	0.1773	17.73
100	–2.1627	0.1150	11.50
120	–2.59524	0.0746	7.46
138.7	–2.9996649	0.0498	4.98 (4 mm)

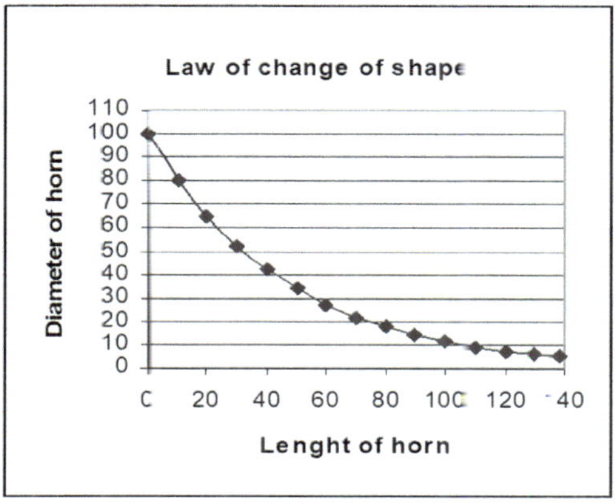

C. **Design of Exponential Concentrator of the Hollow Cylindrical Cross-Section:** Let us consider the case of the exponential horn with a hollow cylindrical cross-section. This type of horn is widely used trepanning that is cutting along the contours of the desired opening (Figure 5.25).

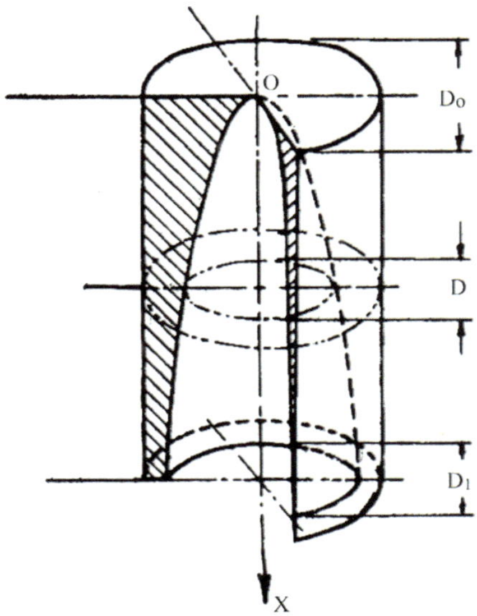

FIGURE 5.25 Exponential horn with a hollow cylindrical cross-section.

K= transformation ratio – amplification ratio

$$K = \sqrt{\frac{(D_o)^2}{(D_a)^2 - (D_1)^2}}$$

where, D_o = diameter of the larger section (outside); D_1 = diameter of the smaller section (inside).

Taper index for an exponential horn:

$$\beta = \frac{\ln(K)}{L}$$

The length of the horn or concentrator ***L*** is usually chosen as half-wave (n=1) or full-wave (n=2).

Length of the exponential horn is given by:

$$L = \frac{n\,C}{f}\sqrt{1 + \left\{\frac{\ln(K)}{2\pi\,n}\right\}^2}$$

And the variation of D is given by:

$D = D_o\sqrt{1-\left(\exp^{-2\beta X}\right)}$ – this is also called Law of change of shape.

A check must be made to ensure that operating frequency exceeds critical frequency f_c given by $f_C = 1.5\frac{\beta C}{2\pi}$.

KEYWORDS

- **electrochemical machining**
- **electro-discharge machining**
- **magnetostriction**
- **magnetostrictive transducer**
- **piezoelectric transducer**
- **ultrasonic machining**
- **velocity transformers**

REFERENCES

1. Benedict, G. F., (1987). *Nontraditional Machining Processes*. New York: Marcel Dekker Inc.
2. Bhattacharayya, A., (1973). *New Technology*. Calcutta: The Institution of Engineers (India).
3. McGeough, J. A., (1988). *Advanced Machining Methods*. London: Chapman and Hall.
4. Pandey, P. C., & Shan, H. S., (1980). *Modern Machining Processes*. New Delhi: Tata McGraw Hill publication Co. Ltd.
5. Mishra, P. K., (1997). *Nonconventional Machining*. Calcutta: Narosa Publication house.
6. Adithan, M., (2009). *Unconventional Machining Processes*. New Delhi: Atlantic Publishers and Distributors.
7. Serope, K., & Steven, R. S., (2007). *Manufacturing Processes for Engineering Material*. Prentice-Hall.
8. Temple, B. J., Ernest, P. D. G., & Ronald, A. K., (2011) Materials and Processes in Manufacturing. John Wiley & Sons.
9. Mikell, P. G., (2010). *Fundamentals of Modern Manufacturing*. John Wiley & Sons.
10. El-Hofy, H., (2005). *Advanced Machining Process*. USA: Tata McGraw Hill.
11. James, A. B., (1991). *Modern Manufacturing Processes*. Industrial Press.

12. Jain, V. K., (2010). *Introduction to Micromachining*. Alpha Science International Limited.
13. Paul, D. J., (2008). *Machining, Fundamentals, and Recent Advances*. Springer Publications.
14. Wit, G., (2008). *Advanced Machining Process of Metallic Materials*. Great Britain: Elsevier.
15. Allesu, K., (2001). *Notes on Manufacturing Science*. ISTE summer school on modern trends in manufacturing science. NIT Calicut.
16. Production Technology, (2008). *Hindustan Machine Tools (HMT)*. New Delhi: Tata McGraw Hill.
17. Nikolaev, G., & Olshansky, N., (1977). *Advanced Welding Processes*. MIR publishers.
18. Ghosh, A., & Mallik, A. K., (2010). *Manufacturing Science.* New Delhi: East-west Press Private Limited.
19. John, A. S., (1987). *Introduction to Manufacturing Processes* (2nd edn.). McGraw Hill Book Company.

CHAPTER 6

Pre-Strain in VHB 4910 Dielectric Elastomer Towards Kinetics of Crystallization

DHANANJAY SAHU,[1] RAJ KUMAR SAHU,[2] and KARALI PATRA[3]

[1] *Senior Research Fellow, Department of Mechanical Engineering, National Institute of Technology, Raipur – 492 010, India*

[2] *Assistant Professor, Department of Mechanical Engineering, National Institute of Technology, Raipur – 492 010, India, E-mail: raj.mit.mech@gmail.com*

[3] *Associate Professor, Department of Mechanical Engineering, Indian Institute Technology, Patna – India*

ABSTRACT

Since two-decades, pre-strain in dielectric elastomers is a persisted proficient technique to improve the performance of soft actuators, sensors, and energy harvesters configuring them. But, information on the perception of pre-strain induced molecular-arrangement and crystallization in the dielectric elastomer is yet to be established. Studies of these structural modifications are important to consider because the change in electromechanical behaviors frequently attributed to chain entanglement for applied pre-strain. Here, the effects of the pre-strain behavior of VHB 4910 dielectric elastomer are deliberated towards electromechanical devices. The influence of pre-strain on electromechanical properties is conferred relating to the phenomenon of macromolecules, chain orientation. The chain entanglement is found to cause stiffening effect and lead strain-induced crystallization. So, the required strain percentage, activation energy, and time of crystallization are estimated based on experimental and analytical work involving the strain rate. These results evidence the kinetics of crystallization in the mentioned elastomer with empirical equivalence, is an inventive contribution towards complete understanding on the effect of pre-strain for actuator application.

6.1 INTRODUCTION

The preliminary introduction of electroactive, dielectric elastomers (DEs) to the research community is specified as electrostrictive polymers to construct micro-actuators [1]. Then the actuation performance of these elastomeric actuators is shown to improve by applying mechanical pre-strain. Different levels of biaxial and uniaxial pre-strain in dielectric elastomers (namely, silicon-based HS3, CF19–2186, and acrylic VHB 4910) are demonstrated to escalate actuation strain beyond a 100% [2]. In the appraisal of silicon, actuators configuring acrylic VHB 4910 elastomer are recognized to perform more efficiently and to be more sensitive to pre-strain. This information fascinated researcher's interest in exploring potential material and technology towards the design and development of soft actuators, sensors, and energy harvesters. Hence, varieties of commercially available polymer were characterized and compared, so, VHB 4910 is recognized as feasible material to configure soft electromechanical devices [3, 4].

Ideally, the DE actuators configure a pre-strained elastomer film mounted on a rigid frame and surfaces coated with a compliant electrode to induct the electric field. On applying electric current, the dielectric behavior of the elastomer keeps the opposite charges separated on the individual surface. These charges tend to attract each other due to Coulomb's law of attraction. This force of attraction induces electrostatic stress-based on the principle of Maxwell stress that compresses the elastomer in the thickness direction, style it to expand laterally. This electromechanical deformation for applied voltage is known as actuation strain and is the measure of actuation performance as related in Eqns. (1) and (2) [2].

$$\sigma_e = \varepsilon\varepsilon_0 E^2 = \varepsilon\varepsilon_0 \left(\frac{V}{d}\right)^2 \tag{1}$$

$$S_Z = -\sigma_e / Y = -\varepsilon\varepsilon_0 E^2 / Y \tag{2}$$

where, σ_e is electrostatic stress, ε is the dielectric permittivity of elastomer, ε_0 is the dielectric permittivity of free-space or vacuum, V is applied voltage, E is the electric field, d is film thickness, and Y is the elastic modulus of the elastomer.

Based on these equations, that is, Eqns. (1) and (2), the electrostatic stress is found depending on the dielectric constant and thickness of the elastomer film, also the actuation strain on both, in addition to the elastic modulus. The ratio of the values for dielectric and elastic behavior of dielectric elastomers is referred to as electromechanical sensitivity, essential to be high for efficient actuation strain [5]. Interestingly, though the dielectric constant and elastic modulus decreases with pre-strain, the actuation performance increases because of drastically lowered elasticity based on the thickness of the elastomer [6].

In the perception of macromolecules, the chain density per unit volume decreases with thickness, which in turn influences material behavior [7]. The monomers known to form macromolecules comprise polarizing dipoles and regulate dielectric behavior under the electric field [8]. In this sense, the decrease in dielectric constant for applied pre-strain is due to distress dipoles within the entangled chain network [9, 10]. So far, these assumptions of chain orientation/entanglement/lengthening/breaking, etc., are without evidence, however, characterization of strain-induced crystallization for VHB 4910 elastomer is perceived as a potential proof, a decade before [11].

Imposing strain-induced crystallization (SIC), act as filler is a very usual prospect to alter electrical, mechanical, thermal, and other physical properties of natural rubber, conjugates [12]. Crystals are usually characterized by X-ray diffraction (XRD) technique to feature its size, degree, etc., and through vibrational spectroscopy to insight the respective change in bond behavior [13]. An alternative approach identifying the elastomer crystallized is analyzing sharp up-turn in the stress-strain curve of the elastomer under tensile load. This is kind to optimize crystallization time and the point of elongation where crystallization exist [14]. These effects of molecular rearrangement and crystallite formation on VHB 4910 dielectric elastomer remain unrevealed. Though its amorphous nature, recognized by XRD is found auspicious for self-healing characteristics and viscoelasticity is explained because of synergistic interaction between covalent/non-covalent bonds [15, 16].

Concerned to research relevance, accurate and precise attempts are given to discuss the effects of pre-strain on the electromechanical and structural behavior of VHB 4910 dielectric elastomer. First of all, it is conversed how important the pre-strain is to improve the actuation performance of actuator made-up of DE. Then the influence of pre-strain on electromechanical behavior is explicated in terms of change in dielectric and mechanical properties, associating the structural adaptation. A sharp upturn in the stress-strain curve is considered as an indicative parameter of strain-induced crystallization; therefore, crystallization time is estimated based on activation energy and strain rate. This work contributes significant information that encourages the study of pre-strain induced adaptation in chain network hence, material behavior for improved actuation strain. Controlled for a certain level, the stiffening of chains in VHB elastomer is found to turns into crystallization, provided desire temperature, and strain rate.

6.2 SIGNIFICANCE OF PRE-STRAIN IN CONFIGURING DE-ACTUATORS

To construct DE-actuator, the elastomer film is pre-strained in any of the regimes (as shown in Figure 6.1) then mounted on a rigid frame, following the coating of a compliant electrode to induct electric current. Hence, electromechanical actuation behavior can be realized by applying the desired voltage, based on the principle of Maxwell stress, as explained earlier.

The pre-strain for DE-actuators is referred to here as the method of tensile loading, (L) of the elastomer for definite deformation to enhance the actuation strain and response time [17, 18]. It helps lower down the required voltage and suppresses the electromechanical pull-in instability, so prevent performance lowering parameters like wrinkling/crumbling at the operational stage [19–25]. This instability in DE-actuators is the failure happens when thickness goes lowers than the threshold and Maxwell stress exceeds the compressive stress in the elastomer. Accordingly, excess pre-stretch must be avoided which is revealed to be 2 for VHB 4910 elastomers, respectively [26]. Beyond this limit, although the pull-in instability gets eliminated, the voltage-induced actuation performance decrease possibly due to an increase in modulus as a result of strain-hardening, stiffening of the elastomer chain [27, 28].

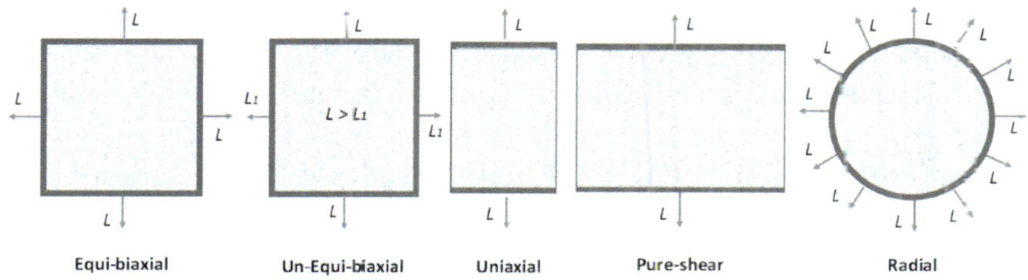

FIGURE 6.1 Regimes of pre-strain for construction of dielectric elastomer actuators, sensors, etc.

Figure 6.2 illustrates the change in areal strain for equi-biaxially pre-strained VHB 4905 (another grade, similar to VHB 4910) elastomer. The electromechanical actuation of a 500% biaxially pre-strained sample is relatively diminished as compared to a sample which is 400% strained and this is reasoned to strain-hardening of elastomer beyond the stretch ratio of 3 [29].

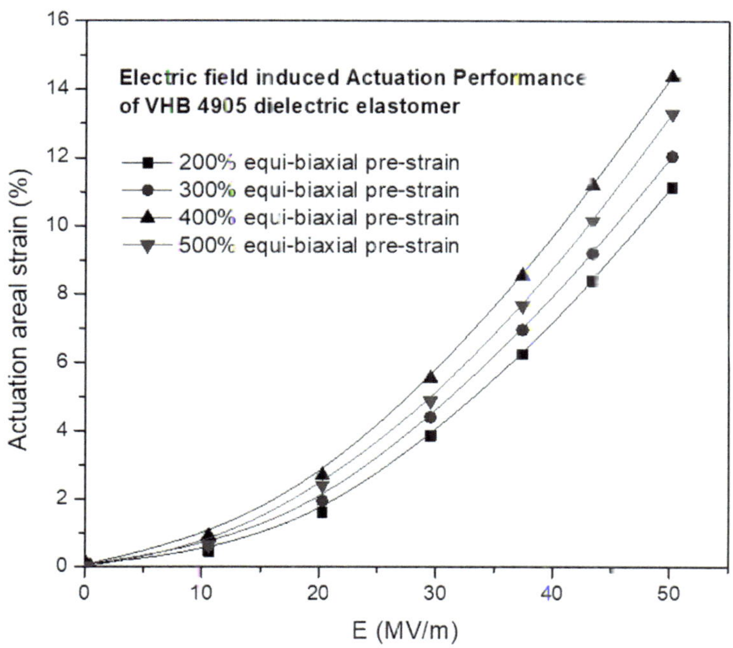

FIGURE 6.2 Electric field (E) induced actuation performance of actuator configuring VHB 4905 dielectric elastomer [29]. Data are represented with permission from Elsevier.

The above discussion indicates the thickness achieved through pre-strain and its consequences on macromolecular chain structure are the elementary reasons to influence actuation strain. Some experimental investigations also assert the improvement in actuation performance is because of a decrease in film thickness independent of the regimes of pre-strain [30, 31]. In the esteem of this, the values of varying thickness of VHB 4910 elastomer for uniaxial and biaxial pre-strain are listed in Table 6.1. In the interim, it is noteworthy that the strain-induced mechanical behavior of the VHB elastomers depends on the strain rate and the elongation for pre-strain [32]. So, there is a possibility that electrical property also. Unfortunately, the dependency of dielectric behavior of VHB 4910 elastomer on pre-strain is still complex to understand based on assumptions of chain orientation/ entanglement [11]. The macromolecular chains itself is a major subject, needs attention on the segmental arrangement (soft/hard/free), degree of polymerization, types of monomers, sorts of a dipole, crosslink density, molecular weight, bond/physical structure, and so on. It needs to understand which macromolecular structure and how the structure controls the electromechanical properties, hence actuation performance.

TABLE 6.1 Pre-Strain Regimes at a Different Level of Deformation and Effective Thickness on VHB 4910 Dielectric Elastomer

Biaxial Pre-Strain (Initial Area 50×50 mm^2) [10]		Uniaxial Strain (Initial Area 20×20 mm^2) [32]	
Pre-stretch ratio (x, y)	**Estimated thickness (μm)**	**Strain % (x)**	**Estimated thickness (μm)**
1×1	1000.0	25	840.0
2×2	250.0	50	760.0
3×3	111.1	10	650.0
4×4	62.5	200	540.0

6.2.1 PRE-STRAIN IN INTERPENETRATION OF DIELECTRIC ELASTOMERS

Interpenetration is the technique of integrating functionalized filler in the inherent structure of elastomer to modify material properties. It is usually performed by pre-molded elastomers like VHBs, smoke sheets, etc. Pre-strain is advanced to interpenetrated VHB 4910 elastomer. The elastomer is biaxially pre-strained for 400% and sprayed with filler dispersed crosslinking reagent, followed by free radical polymerization. In doing so, viscoelastic behavior is claimed to condense so mechanical energy density and response time were improved [33, 34]. In this way, the trimethylolpropane trimethylacrylate (crosslinking reagent) is found as a potential filler to improve electromechanical sensitivity of VHB 4910 elastomer [7]. Similarly, the use of plasticizers namely, dioctyl phthalate and dibutyl phthalate are found appropriate fillers to decrease the elastic modulus of acrylate rubber [35]. Some other filler is known to simultaneously influence the mechanical, thermal, and electrical properties include barium titanate, graphene oxide, vegetable starch, etc., (discussion on which is out-of-scope of this article) [36–39].

Despite the interpenetration is recognized as a suitable technique to improve the electromechanical behavior of dielectric elastomers. The consequences of applied pre-strain like stress concentration at filler-matrix interface, crack nucleation/propagation, change in dielectric behavior, energy density, etc., are yet to be considered in a similar investigation.

6.3 INFLUENCES OF PRE-STRAIN ON MACROMOLECULAR CHAINS OF DES

In general, the most desired dielectric and elastic behavior of DEs decreases with the application of pre-strain and relevantly attributed to influence on-chain entanglement and bond structure, in recent work [9, 40–42]. To elucidate these phenomenal possessions, studies regarding macromolecules parameters including monomers, polarizing dipoles, degree of polymerization, crosslink density, molecular weight, etc., are considered here.

Figure 6.3 shows the monomer's linkage that forms a macromolecular chain network and their entangling behavior. On stretching the adjacent chains, come closer from the habitual position, so give rise to the chances to lose the intermolecular bond strength or to increase bond length. The disordered monomer comprises the distressed polarizing dipoles which in turn affect dielectric, other bond-related behavior of VHB 4910 elastomer, expected to cause high electrical dissipation in form of heat [16, 43, 44]. In this sense, entangling of macromolecular chains should be prevented or the density of monomers can be increased. The escalation of monomer, in context with crosslink density, is recently proved to improve electromechanical sensitivity. It is revealed that the addition crosslinking monomer bridge the gap between parent monomers, so increases the dielectric constant, develops a tendency to prevent chain entanglement [7]. Fundamental relation of chain density with dielectric constant and elastic modulus are presented in Eqns. (3) and (4) [45, 46];

$$\varepsilon = 0.57v^2 - 1.22v + 1.92 \tag{3}$$

$$Y = 3\nu kT \tag{4}$$

where, ν is the crosslink density, k is Boltzmann constant, and T is the absolute temperature.

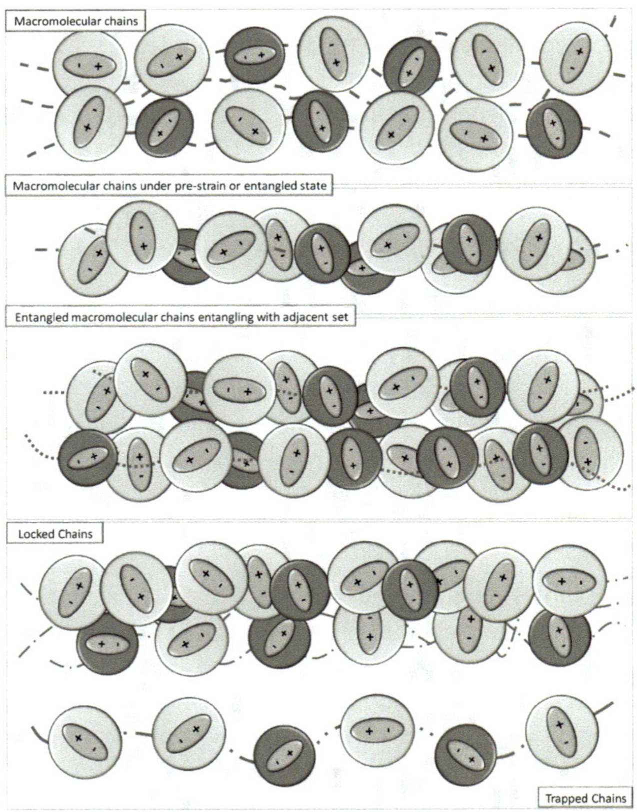

FIGURE 6.3 Entangling in the macromolecular chain of dielectric elastomer due to applied pre-strain.

In these terms, the pre-strain induced mobility of ether dipoles are said to control dielectric constant and the presence of the chalcogen group is revealed to govern electric behavior [29, 47]. Further, an innovative impression regards the effect of applied strain on a network of silicon-DEs is; the entangled chains act as new crosslinks with more dynamic behavior. Such crosslinks are said to slide over each other and/or to disentangle. The disentangled chain under loading condition can be referred to as "trapped" and those cannot as "locked" chains that constantly contributing the elasticity to elastomer [48]. These attributions are based on experimental results of natural rubber (NR) that also suggests the stiffened entanglement, strain-hardening lead strain-induced crystallization [49]. From the discussion of strain-hardening, a sudden upturn in the stress-strain (SN) curve of NR is endorsed as the data point where stiffening occurs [50]. Noticeably, this occurs much prior to biaxial deformation than the uniaxial way. This sharp upturn of the SN curve is also referred to as an indicative parameter for the kinetics of crystallization in PEEK [14].

In the case of VHB 4910 elastomer, though the entangling and stiffening of chains are frequently attributed to pre-strain in the available literature, characterization of SIC has remained unrevealed. Here, we investigated the aspects of crystallization by analyzing the SN curve for tensile loading. Figure 6.4 shows the mechanical behavior of VHB 4910 under tensile loading at different elongation rates, performed on individual specimens of 6 × 1 mm^2 cross-sectional area and 50 mm length. The stress-strain curves are examined for sudden sharp turns found at (797.68, 299.73), (781.56, 308.9), (740.98, 295.77) for 100, 200, and 300 mm/min elongation rate, respectively. Based on these data, the corresponding time for stiffening (crystallization) in elastomer are listed in Table 6.2.

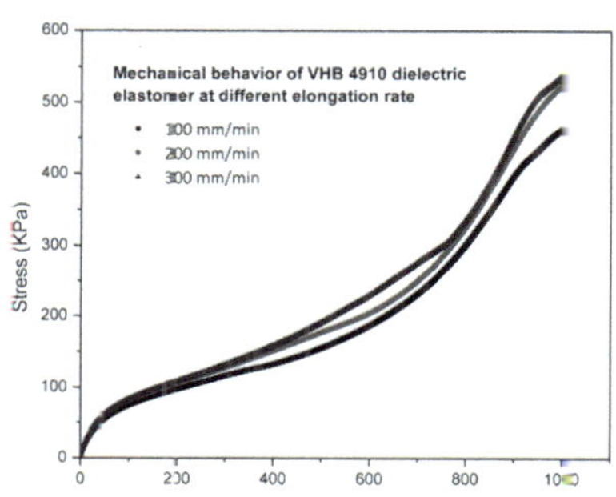

FIGURE 6.4 Stress-strain curve of VHB 4910 dielectric elastomer under tensile load at a different elongation rate.

TABLE 6.2 Detail of the Sudden Upturn in the Stress-Strain Curve of VHB 4910 Elastomer Under Static Load

Elongation Rate (mm/ min)	Elongation Rate (mm/sec)	Strain Rate (per sec)	Turn at Strain Percentage (%)	Crystallization Time (sec)
100	1.666	0.0328	797.68	243.195
200	3.3333	0.0656	781.56	119.14
300	5	0.098	740.98	75.61

The values in Table 6.1 are evaluated based on Eqns. (3) and (4);

$$\text{Strain rate} = \frac{\text{Elongation rate}}{\text{Initial length of specimen}} \tag{3}$$

$$\text{Crystallization time} = \frac{\text{Strain}}{\text{Strain rate}} \tag{4}$$

These results revealed the stiffening behavior of VHB 4910 elastomer has a tendency to go through kinetics of crystallization like the other several other elastomers and it increases with strain rate as shown in Figure 6.5.

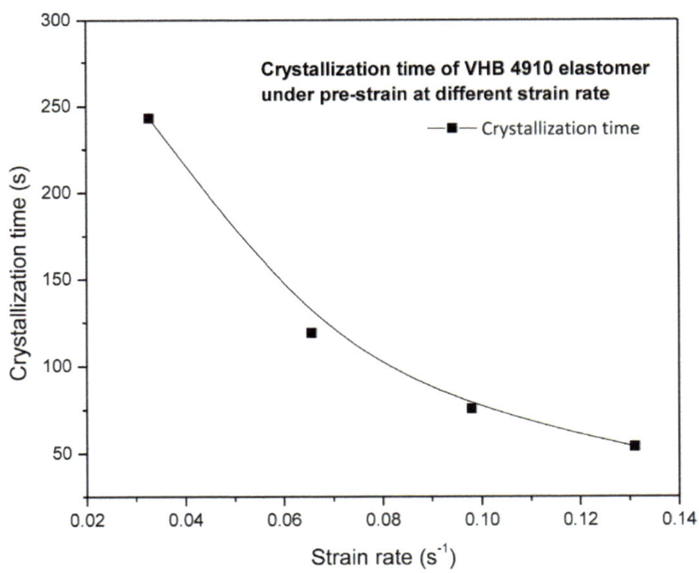

FIGURE 6.5 Crystallization time for VHB 4910 elastomer under tensile load at the varying strain rate.

6.4 ACTIVATION ENERGY: CALCULATION OF CRYSTALLIZATION TIME

Activation energy from point of SIC is the minimum energy required to start the crystallization process under tensile load. The energy is supplied to the elastomer from tensile stress generated nucleation against deformation. Yeh et al. [51] studied the empirical relation shown in Eq. (5) to relate exponential dependency of the reciprocal of induction time, t_i on polyethylene (PE).

$$\frac{1}{ti} = \frac{\gamma e^{\frac{Ea}{RT}}}{A} \tag{5}$$

where, t_i is induction time, A is constant, E_a is the activation energy, R is the universal gas constant, Υ is strain rate.

It is observed that the induction time of crystallization is dependent on activation energy (E_a). This energy can be calculated based on thermogravimetric (TGA) data as studied by Flynn and Wall [52]. Accordingly,

$$Ea = -4.35d(\frac{\ln \beta}{1/T}) \tag{6}$$

where, β is the heating rate, and T is crystallization temperature.

The TGA curve of VHB 4910 is referred here to the work of Sheng et al. [44], performed at a heating rate (β) of 10°C/min. The sharp upturn is observed at a weight percentage of 65 and the corresponding temperature is noted 641 K. Similarly, temperatures are assumed based on trial and error method for different heating rates are shown in Table 6.3. Using Eq. (6), activation energy is calculated for VHB 4910 and it comes out to be 45327 cal/mol.

TABLE 6.3 Data for Calculating Activation Energy of Crystallization in VHB 4910 Dielectric Elastomer

Temperature (K)	β (°C/sec)	1/T	Ln(β)
641	0.16	0.00156	–1.83258
658	0.25	0.00152	–1.38629
671	0.33	0.00149	–1.10866
684	0.40	0.00146	–0.9163

6.4.1 CRYSTALLIZATION TIME CALCULATION

For crystallization time calculation, empirical Eq. (5) is used. Here, based on referred literature constant A is proposed as 5.8×10^{16}

$\gamma = 0.098\ s^{-1}$
R = 1.987 cal/mole-K
T = 623 K
$E_a = 45327$ cal/mole
$t_i = 74.14$ seconds

Thus, the crystallization time for a strain rate of 0.0098 s^{-1} is evaluated to be 74.14 seconds which is expected in the range of milliseconds for strain rates between 8 s^{-1} to 50 s^{-1}. These results show the trend that at lower strain rates molecular interactions will take place slowly thus resulting in larger crystallization times. However, some error is also estimated comparing the values obtained for crystallization time based on a sharp upturn in SN curve, certainly because of made assumptions and inadequate information for VHB.

6.4.2 COMPARISON OF RESULTS

The value obtained for crystallization time and errors from sharp turn calculations and analytical models are represented in Table 6.4 and plotted in Figure 6.6.

TABLE 6.4 Error in Experimental and Model Crystallization Time in VHB 4910 Dielectric Elastomer

Analytical Model (s)	Sharp Upturn Theory (s)	Strain Rate (mm/s)	Error (%)
55.46	53.86	0.131	2.88
74.14	75.61	0.098	1.94
135.57	119.14	0.0656	12.11
221.52	243.15	0.0328	8.89

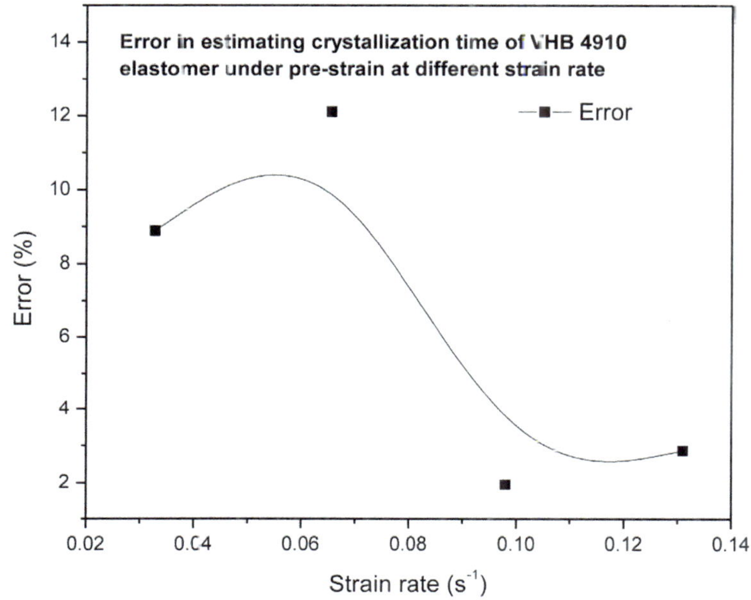

FIGURE 6.6 Error in the estimation of the crystallization time for VHB 4910 elastomer.

Figure 6.6 shows the error % increases to a maximum value then decreases significantly. Thus, moderate strain rates produce a maximum error in the comparison of an analytical and theoretical model.

6.5 SUMMARY

In this preliminary investigation of pre-strain induced crystallization in VHB 4910 dielectric elastomer, the entangling and stiffening behavior of macromolecular chains is highlighted as a crucial aspect in regards to attributions in the available literature. Pre-strain is explained to decrease film thickness and suppress pull-in instability to a certain limit depends on the stiffening of elastomer chains. A new perception on-chain entangling is exposed in the form of trapped chain and locked chains that contributed electric and elastic behavior. Stiffening of locked chains beyond stretch ratio 3 is revealed to conduct kinetics of crystallization in the subjected elastomer, so influences material properties. The strain percentage, time, and activation energy required for possession of crystallization are estimated and found to decrease with increasing strain rate. These results evidenced significant information to persuade in-situ study for characterizing the kinetics of crystallization in dielectric elastomer using XRD techniques.

ACKNOWLEDGMENT

Authors acknowledge the help rendered by the Department of Science and Technology, SERB India-New Delhi, in the form of the fund through grant file no. ECR/2016/000581.

KEYWORDS

- **crystallization**
- **dielectric elastomers**
- **electromechanical properties**
- **macromolecules**
- **micro-actuators**
- **stress-strain**

REFERENCES

1. Pelrine, R., Kornbluh, R., Joseph, J., Chiba, S., & Park, M., (1997). Electrostriction of polymer films for microactuators. In: *Proc. IEEE Tenth Annu. International Work, Micro Electro-Mech. Syst.* (pp. 238–243). Nagoya, Japan.
2. Pelrine, R., Kornbluh, R., Pei, Q., & Joseph, J., (2000). High-speed electrically actuated elastomers with strain greater than 100%. *Science, 80, 287*(5454), 836–840. doi: 10.1126/science.287.5454.836.
3. Bar-Cohen, Y., & Zhang, Q., (2008). Electroactive polymer actuators and sensors. *MRS Bull., 33*, 173–181. doi: 10.1557/mrs2008.42.
4. Bar-Cohen, Y., & Anderson, I. A., (2019). Electroactive polymer (EAP) actuators-background review. *Mech. Soft Mater*, *1*(5), 1–14. doi: 0.1007/s42558-019-0005-1.
5. Ning, N., Li, S., Sun, H., Wang, Y., Liu, S., Yao, Y., Yan, B., Zhang, L., & Tian, M., (2017). Largely improved electromechanical properties of thermoplastic polyurethane dielectric elastomers by the synergistic effect of polyethylene glycol and partially reduced graphene oxide. *Compos. Sci. Technol., 142*, 311–320. doi: 10.1016/j.compscitech.2017.02.015.

6. Huang, J., Shian, S., Diebold, R. M., Suo, Z., & Clarke D. R., (2012). The thickness and stretch dependence of the electrical breakdown strength of an acrylic dielectric elastomer. *Appl. Phys. Lett., 101*, 1–4. doi: 10.1063/1.4754549.
7. Sahu, D., Sahu, R. K., & Patra, K., (2019). Effects of crosslink density on the behavior of VHB 4910 dielectric elastomer. *J. Macromol. Sci. Part a Pure Appl. Chem., 56*, 821–829. doi: 10.1080/10601325.2019.1610329.
8. Sahu, R. K., Yadu, S., Singh, V., Raja, S., & Patra, K. (2017). The effect of micro molecular parameters on the actuation performance of electro active polymers. In: *Int. Mech. Eng. Congr. Expo* (pp. 71272, 1–6). ASME. doi: 10.1115/IMECE2017–71272.
9. Zhu, J., & Luo, J., (2017). Effect of entanglements on the electromechanical stability of dielectric elastomers. *Epl., 119*. doi: 10.1209/0295–5075/119/26003.
10. Vu-cong, T., Ngyen-Thi, N., & Sylvestre, A., (2014). How does static stretching decrease the dielectric constant of VHB 4910 elastomer? In: *Electroact. Polym. Actuators Devices (EAPAD)* (pp. 1–9). SPIE. doi: 10.1117/12.2045042.
11. Jean-Mistral, C., Sylvestre, A., Basrour, S., & Chaillout, J. J., (2010). Dielectric properties of polyacrylate thick films used in sensors and actuators. *Smart Mater. Struct*., 191–199. doi: 10.1088/0964–1726/19/7/075019
12. Lizundia, E., Larran, A., Larrañaga, A., & Lizundia, E., (2018). *Strain-Induced Crystallization*. doi: 10.1016/B978–0-12–809453–2.00015–3.
13. Young, R. J., & Eichhorn, S. J., (2007). Deformation mechanisms in polymer fibers and nanocomposite. *Polymer (Guildf), 48*, 2–18. doi: 10.1016/j.polymer.2006.11.016.
14. Weiss, R. A., (1988). Strain-induced crystallization behavior of poly (ether ether ketone) (PEEK). *Polym. Engineering Science, 28*, 6–12, doi: 10.1002/pen.760280103.
15. Fan, F., & Szpunar, J., (2015). The self-healing mechanism of an industrial acrylic elastomer. *J. Appl. Polym. Sci., 132*. doi: 10.1002/app.42135.
16. Fan, F., & Szpunar, J., (2015). Characterization of visco elasticity and self-healing ability of VHB 4910. *Macromol. Mater. Eng., 300*, 99–106. doi: 10.1002/mame.201400122.
17. Newell, B., Krutz, G., Stewart, F., & Pascal, K., (2016). Novel pre-strain method for dielectric electroactive polymers. *Proc. SPIE 9798. Electroact. Polym. Actuators Devices, 9798*, 979837, 1–7. doi: 10.1117/12.2220098.
18. Saini, A., Ahmad, D., & Patra, K., (2016). Electromechanical performance analysis of inflated dielectric elastomer membrane for micro pump applications. *Proc. SPIE 9798 Electroact. Polym. Actuators Devices, 9798*, 979813, 1–7. doi: 10.1117/12.2219032
19. Sakorikar, T., Kavitha, M. K., Vayalamkuzhi, P., & Jaiswal, M., (2017). Thickness-dependent crack propagation in uniaxially strained conducting graphene oxide films on flexible substrates. *Sci. Rep., 7*, 1–10. doi: 10.1038/s41598–017–02703–2.
20. Zhao, X., & Wang, Q., (2014). Harnessing large deformation and instabilities of soft dielectrics : Theory, experiment, and application. *Appl. Phys. Rev., 1*, 1–24. doi: 10.1063/1.4871696.
21. Liu, X., Road, X. W., Chen, H., & Li, B., (2015). Experimental study on electromechanical failure of dielectric elastomer actuator. In: *Int. Mech. Eng. Congr. Expo* (pp. IMECE2015-50406, V009T12A079, 1–7). doi: 10.1115/IMECE2015-50406.
22. Li, K., (2018). Voltage-induced wrinkling in a constrained annular dielectric elastomer film. *J. Appl. Mech.-ASME, 85*, 1–10. doi: 10.1115/1.4038427.

23. Dubowsky, S., (2006). Large-scale failure modes of dielectric elastomer actuators. *Int. J. Solids Struct., 43*, 7727–7751. doi: 10.1016/j.ijsolstr.2006.03.026.
24. Lai, W., & Lai, W., (2011). Characteristics of dielectric elastomers and fabrication of dielectric elastomer actuators for artificial muscle applications. *Graduate Theses and Dissertations*, p. 12183. https://lib.dr.iastate.edu/etd/12183 (accessed on 24 September 2020).
25. Liwen, H., Lou, J., Du, J., & Wang, J., (2017). Finite bending of a dielectric elastomer actuator and pre-stretch effects. *Int. J. Mech. Sci., 122*, 120–128. doi: 10.1016/j.ijmecsci.2017.01.019.
26. Jiang, L., Betts, A., Kennedy, D., & Jerrams, S., (2016). Eliminating electromechanical instability in dielectric elastomers by employing pre-stretch. *J. Phys. D. Appl. Phys., 49*, 1–11. doi: 10.3390/act7020032.
27. Zhu, J., Cai, S., & Suo, Z., (2010). Resonant behavior of a membrane of a dielectric elastomer. *Int. J. Solids Struct., 47*, 3254–3262. doi: 10.1016/j.ijsolstr.2010.08.008.
28. Koh, S. J. A., Keplinger, C., Li, T., Bauer, S., & Suo, Z., (2011). Dielectric elastomer Generators: How much energy can be converted? *IEEE/ASME Trans. Mechatronics, 16*, 33–41. doi: 10.1109/TMECH.2010.2089635.
29. Liu, L., Huang, Y., Zhang, Y., & Allahyarov, E., (2018). Understanding reversible Maxwellian electro actuation in a 3M VHB dielectric elastomer with restraint. *Polymer (Guild), 144*, 150–158. doi: 10.1016/j.polymer.2018.04.048.
30. Kumar, A., Ahmad, D., & Patra, K., (2018). Dependence of actuation strain of dielectric elastomer on equi-biaxial, pure shear and uniaxial modes of pre-stretching. *IOP Conf. Ser. Mater. Sci. Eng., 310*, 1–9. doi: 10.1088/1757–899X/310/1/012104.
31. Zhao, Y., Zha, J. W., Yin, L. J., Gao, Z. S., Wen, Y. Q., & Dang, Z. M., (2018). Remarkable electrically actuation performance in advanced acrylic-based dielectric elastomers without pre-strain at very low driving electric field. *Polymer (Guildf), 137*, 269–275. doi: 10.1016/j.polymer.2017.12.065.
32. Sahu, R. K., & Patra, K., (2016). Rate-dependent mechanical behavior of VHB (4910). Elastomer. *Mech. Adv. Mater. Struct., 23*, 170–179. doi: 10.1080/15376494.2014.949923.
33. Ha, S. M., Park, I. S., Wissler, M., Pelrine, R., Stanford, S., Kim, K. J., Kovacs, G., & Pei, Q., (2008). High electromechanical performance of electro elastomers based on interpenetrating polymer networks. In: *Electroact. Polym. Actuators Devices (EAPAD)* (p. 69272C). SPIE. doi: 10.1117/12.778282.
34. Ha, S. M., Yuan, W., Pei, Q., Pelrine, R., & Stanford, S., (2007). Interpenetrating networks of elastomers exhibiting 300% electrically-induced area strain. *Smart Mater. Struct., 16*, S280–S287. doi: 10.1088/0964–1726/16/2/S12.
35. Jha, A., Dutta, B., & Bhowmick, A. K., (1999). Effect of fillers and plasticizers on the performance of novel heat and oil-resistant thermoplastic elastomers from nylon-6 and acrylate rubber blends. *J. Appl. Polym. Sci., 74*, 1490–1501. doi: 10.1002/(SICI)1097–4628(19991107)74:6<1490:AID-APP22>3.0.CO;2-U.
36. Ziegmann, A., & Schubert, D. W., (2018). Influence of the particle size and the filling degree of barium titanate filled silicone elastomers used as potential dielectric elastomers on the mechanical properties and the cross-linking density. *Mater. Today Commun., 14*, 90–98. doi: 10.1016/j.mtcomm.2017.12.013.

37. Qiu, Y., Wang, M., Zhang, W., Liu, Y., Li, Y. V., & Pan, K., (2018). An asymmetric graphene oxide film for developing moisture actuators. *Nanoscale, 10*, 14060–14066. doi: 10.1039/c8nr01785a.
38. Cao, C., Feng, Y., Zang, J., López, G. P., & Zhao, X., (2015). Tunable lotus-leaf and rose-petal effects via graphene paper origami. *Extrem. Mech. Lett., 4*, 18–25. doi: 10.1016/j.eml.2015.07.006.
39. Vuorinen, J., Sarlin, E., Das, A., Poikelispää, M., & Shakun, A., (2017). Vegetable fillers for electric stimuli responsive elastomers. *J. Appl. Polym. Sci. 134*, 45081. doi: 10.1002/app.45081.
40. Javadi, S., Panahi-Sarmad, M., & Razzaghi-Kashani, M., (2018). Interfacial and dielectric behavior of polymer nano-composites: Effects of chain stiffness and cohesive energy density. *Polym. (United Kingdom), 145*, 31–40. doi: 10.1016/j.polymer.2018.04.061.
41. Zhang, Q. P., Liu, J. H., Liu, H. D., Jia, F., Zhou, Y. L., & Zheng, J., (2017). Tailoring chain length and cross-link density in dielectric elastomer toward enhanced actuation strain. *Appl. Phys. Lett., 111*, 1–4. doi: 10.1063/1.5001666.
42. Gu, G., Zhu, J., Zhu, L., & Zhu, X., (2017). A survey on dielectric elastomer actuators for soft robots. *Bioinspir. Biomim., 12*, 1–22.
43. Jean-Mistral, C., (2012). Impact of the nature of the compliant electrodes on the dielectric constant of acrylic and silicone electroactive polymers. *Smart Mater. Struct., 21*, 1–10. doi: 10.1088/0964–1726/21/10/105036.
44. Sheng, J., Chen, H., Qiang, J., Li, B., & Wang, Y., (2012). Thermal, mechanical, and dielectric properties of a dielectric elastomer for actuator applications. *J. Macromol. Sci. Part B., 51*, 2093–2104. doi: 10.1080/00222348.2012.659617.
45. Turasan, H., Barber, E. A., Malm, M., & Kokini, J. L., (2018). Mechanical and spectroscopic characterization of cross-linked zein films cast from solutions of acetic acid leading to a new mechanism for the cross-linking of oleic acid plasticized zein films. *Food Res. Int., 108*, 357–367. doi: 10.1016/j.foodres.2018.03.063.
46. Chua, J., & Tu, Q., (2018). A molecular dynamics study of cross-linked phthalonitrile polymers: The effect of crosslink density on thermo mechanical and dielectric properties. *Polymers (Basel), 10*, 1–11. doi: 10.3390/polym10010064.
47. Sahu, D., Sahu, R. K., & Patra, K., (2019). Effects of uniaxial and biaxial strain on molecular structure of VHB 4910 dielectric elastomer. In: *Adv. Polym. Compos. Mech. Charact. Appl.* (pp. 1–7). doi: 10.1063/1.5085603.
48. Mazurek, P., Vudayagiri, S., & Skov, A. L., (2019). How to tailor flexible silicone elastomers with mechanical integrity : A tutorial review. *Chem. Soc. Rev.* doi: 10.1039/c8cs00963e.
49. Toki, S., Che, J., Rong, L., Hsiao, B. S., Amnuaypornsri, S., Nimpaiboon, A., & Sakdapipanich, J., (2013). Entanglements and networks to strain-induced crystallization and stress-strain relations in natural rubber and synthetic polyisoprene at various temperatures. *Macromolecules, 46*, 5238–5248. doi: 10.1021/ma400504k.
50. Davidson, J. D., & Goulbourne, N. C., (2013). A nonaffine network model for elastomers undergoing finite deformations. *J. Mech. Phys. Solids, 61*, 1784–1797. doi: 10.1016/j.jmps.2013.03.009.
51. Yeh, G. S., Hong, K. Z., & Krueger, D. L., (1979). Strain-induced crystallization, Part IV: Induction time analysis. *Polym. Eng. Science, 19*, 401–405.

52. Flynn, J. H., & Wall, L. A., (1966). A quick, direct method for the determination of activation energy from' thermo gravimetric data. *Polym. Lett., 4*, 323–328. doi: 10.1098/rstb.1988.0133.

CHAPTER 7

End Life Cycle Recycling Policy Framework for Commercially Available Solar Photovoltaic Modules and Their Environmental Impacts

MANISHA SHEORAN,[1] PANCHAM KUMAR,[2] and SUSHEELA SHARMA[1]

[1]*Department of Basic Science, Bhartiya Skill Development University Jaipur, Rajasthan, India*

[2]*School of Electrical Skills, Bhartiya Skill Development University Jaipur, Rajasthan, India*

ABSTRACT

Increasing changes in weather conditions like up-surged global temperature, pollution, population explosion, and economic backwardness being the major problems faced by our society in the present century. The natural resources have been exploited to their peek by the increasing population. Emerging industries and enhancement of lifestyle of people across the globe have further exhausted the conventional sources of energy with mother Earth being in its usual form. The growing taste of urbanization and modernization in developing countries is causing major repercussions for the environment and energy sector. The requirement of energy and its associated amenities indulges humankind into the blooming of social, economic, and health benefits. To meet these requirements the necessity of renewable energy sources was emphasized. With the ubiquitous presence of the Sun, solar energy proved to be an inexpensive and versatile source of energy. A further provision of government subsidies the solar photovoltaic power bloomed across the world. Using the life cycle assessment (LCA) approach, an in-depth study of the various inputs and outputs in a solar PV system is studied. Though with increasing agitation in climate perturbed a sustainable and greener route to decrease the greenhouse gaseous emissions is mandatory. The use of renewable sources of energy in a clever mode will be bringing drastic changes in the improvement of energy issues in a greener way. The clean and green way of renewable energy resources has to be applied in every niche of the world. Solar energy proved to be a major contributor to this task. The governmental initiative of more PV installations is leading to a macroscopic proliferation of solar photovoltaic waste accumulation in the country, which has become the utmost issue of the hour to be handled by PV scarp management and recycling policy. Existing PV technologies will be evaluated and their effects on the environment, human health, social and economic aspects will be analyzed in depth. Major PV module recycling methods are evaluated and the material recovery in economic terms is recorded. With the sky rocking PV scrap, India is obligate for inclination towards the PV recycling policy framework. The main focus of the present work is towards the effective remedy of the environmental and socio-economic impact arising in the life cycle assessment of c-Si, CdTe, CdS, and CIGS. A sustainable policy is to be put forward to potentially tackle the upcoming problems from the PV waste generation and accumulation.

7.1 INTRODUCTION

Ever since the homo sapiens came on this planet they have been using energy in one form or the other. For the very time, they used fire originated from rubbing of stones together to cook the flesh. But with the passage of time, the process of manipulating the sources of energy came into emergence. To further meet the demands of food cultivation of land was initiated and further energy was generated from water, wind, and Sun. The use of fossil fuels is increasing the environmental problems like acid rain, air pollution, and global warming. To avoid these environmental problems a migration towards the sustainable energy resource is required and solar PV energy has the capability to full fill this need. PV converts the solar energy directly from the sun into the electric energy by the photovoltaic effect. During the operational phase of the solar photovoltaic system, there is no utilization of the conventional energy sources and there is no emission of greenhouse gases, hence this technology is guided as environmentally friendly.

With the growing population, the energy requirements of the world are continuously increasing. Urbanization further added to this demand. Energy from the solar radiation can be converted into consumable form by the use of various tools and techniques. Mainly three techniques can be applied to fulfill the solar energy conversion:

1. Solar energy can be converted into electricity by using photovoltaic technology.
2. Solar energy can be converted into heat energy by using solar thermal technology.
3. Solar energy can be directly transmitted to buildings using non-mechanical methods.

Environmental effect of the photovoltaic panels is divided into two categories:

1. Advantageous impact; and
2. Disadvantageous impact.

The advantageous impact is assessed by the use of photovoltaic technology to combat the harmful effects of the conventional resources on the environment like pollution, increased global temperature. The disadvantageous impact is due to the decommissioning and manufacturing of the photovoltaic panels. Increasing energy demands all over the world have led to the discovery of new energy sources after the uncontrolled use of conventional energy sources. Solar photovoltaic technology has emerged as the most promising technology to improve the energy security and to mitigate the climate changes. PV technology has emerged as a clean energy source, so its proper end of life treatment is a concerning issue migration towards an environmentally friendly resource from the conventional resource is the key step for the putting an end to the greenhouse gaseous emissions.

7.2 METHODOLOGY

The total installed capacity at the end of 2018 was 500 GW globally, by 2023, the global PV installations will reach to 1296 GW as mentioned in IEA- PVPS 2019, and further, by the end of 2050, it will reach to 4500

GW as quoted in IRENA 2016. India will attain a total installation of 28 GW as on March 2019 as mentioned in IRENA. With the increasing overexploitation of conventional energy sources and degradation of the earth's life support system, many industries and businesses have started to assess their product's environmental impact. One such tool used for this is LCA, i.e., Life cycle assessment. A life cycle assessment is performed to understand the cumulative input and output [1, 2].

LCA is a tool to assess the environmental impacts and resources used throughout a product's life cycle and consider all attributes or aspects of the natural environment, human health, and resources and can be defined as a method for analyzing and assessing the environmental impacts of a material, product, or service along its entire life cycle (ISO-2005). Thus, ISO14040 defined LCA as the "compilation and evaluation of the inputs, outputs, and potential environmental impacts of a product system throughout its life cycle" (ISO-2006). For further classification of it into environmentally friendly technology, the whole life cycle assessment of the PV technology is taken into analysis. The utilization of solar energy came out to be a very alluring option. With the increasing population and economy of the world, the traditional sources of energy have failed to fulfill the demands of energy. And the greenhouse gaseous emissions have also increased and this is causing environmental deterioration. Therefore, a sustainable energy resource is strictly required to accomplish the needs of energy and to combat the environmental deteriorations [3, 4]. Various types of harnessing systems are developed for solar energy like building integrated PV and building applied this includes facade, sloped or pitched, submerged PV, floating PV, solar tree, etc. Among these solar PV trees are eventually being trendy for the generation of electricity because of lesser requirements of land area. SPTV are being more prominent in urban areas due to the lesser availability of space. Moreover, these resemble a natural tree [5].

The complete life cycle involves the manufacturing steps, assemblage steps, processing of components of a photovoltaic system, transportation of required materials, installation, and furnishing of the solar PV system, and finally the disposal after the complete utilization of the product [6, 7]. Consequently, the life cycle assessment is brought out. Through this, an exact and accurate study is shaped which considers all aspects of the environment from manufacturing to its payback time of energy.

LCA is defined as the accumulation and interpretation of the inputs required and the outcome of the technology during the overall lifetime. LCA is comprised of four areas:

1. Design and outlook;
2. Reservoir assay; and
3. Effect/impact analysis.

On the basis of the above areas, an entire outline of the LCA can be shaped (Figure 7.1).

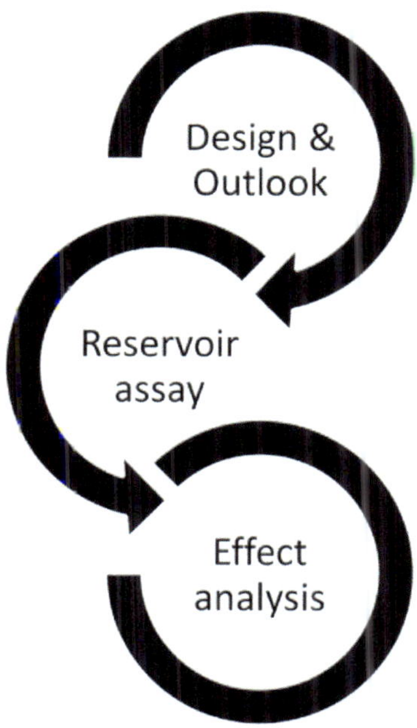

FIGURE 7.1 Outline of the life cycle assessment.

The design and outlook of LCA consider the manufacturing stage means from cradle to gate and end life stage means cradle to grave. Hence, LCA speculates the entire stages of the life cycle of the product. All the stages from raw material acquisition to the assemblage process to the operational phase are all veiled into the life cycle assessment of the PV technology. Reservoir assay deals about the raw material extraction for the input [8, 9]. It also includes the cost inclusion of all the input steps and the output steps of the life cycle of photovoltaic panels; transportation cost is also included here. Effect analysis inculcates the entire assessment of the photovoltaic panels. All the inputs from the surroundings and the output going to the various other sources like air, water, land are analyzed by the approach of raw material flow. Climate change, human toxicity, depletion of ozone, eutrophication, depletion of water resources, acidification, etc., are all included under the effect analysis [10, 11].

Electricity producing systems based on photo-voltaic phenomenon impart lower emissions of greenhouse gases in comparison to systems generating energy from the conventional sources. If this energy generation based on the PV system is attached to the grid than the greenhouse emissions will be high [12]. For further analysis of the environmental impacts caused by the electricity generation, LCA will be deeply studied and EPBT is taken as the index.

Air emission, water use, and land use are considered as the indicators of the life cycle. All over, the environmental attributes are examined. LCA analyzes the environmental burden of products from cradle to grave, from raw material acquisition, manufacturing of the product, use, and maintenance of the product and finally to its end of life management, either by reuse, recycle or reducing [13]. Input is analyzed in terms of the amount of energy, the energy payback time is also given which ensures that solar energy is a clean energy source and the operation period is assumed to be of 27–30 years, hence it is more environmentally friendly than fossil fuels.

The life cycle stages of solar PV are depicted in Figure 7.2.

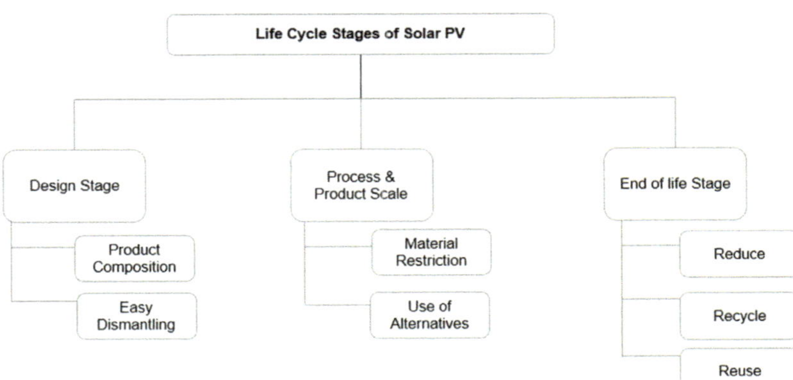

FIGURE 7.2 Life cycle stages of solar PV.

A photovoltaic module consists of an assembly of solar cells packed into a protective multilayer structure consisting of a front cover, electrical circuit interconnecting solar cells, encapsulant layer, backsheet, and metal frames for support [14]. Major share is of glass and aluminum. With the growing PV technologies, the use of metals like Cd, Te, Sn, In, Ga, Se is increasing and they possess toxicity effects [15, 16]. Most materials other than glass and aluminum are usually recovered in less quantity [17]. Some existing PV recycling methods are shown in Figures 7.3–7.5.

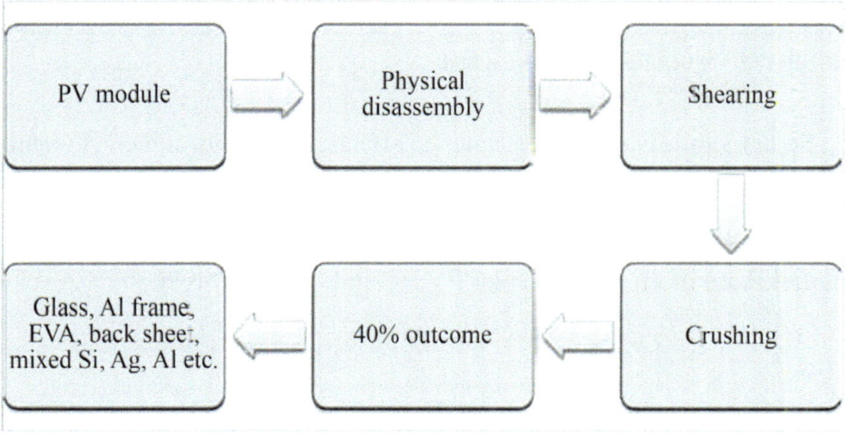

FIGURE 7.3 Mechanical recycling method.

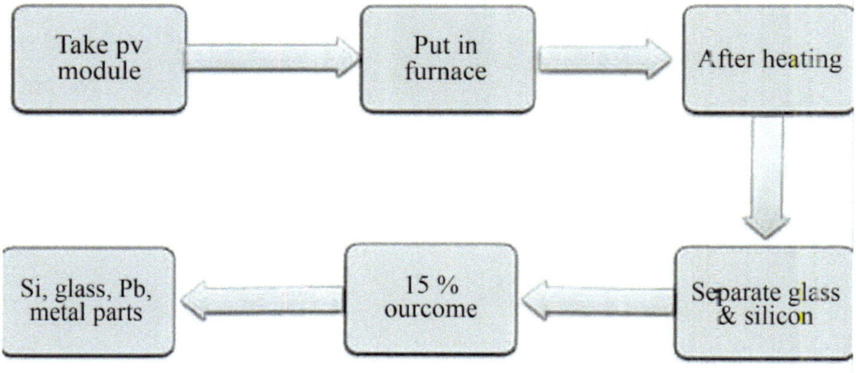

FIGURE 7.4 Thermal method of recycling.

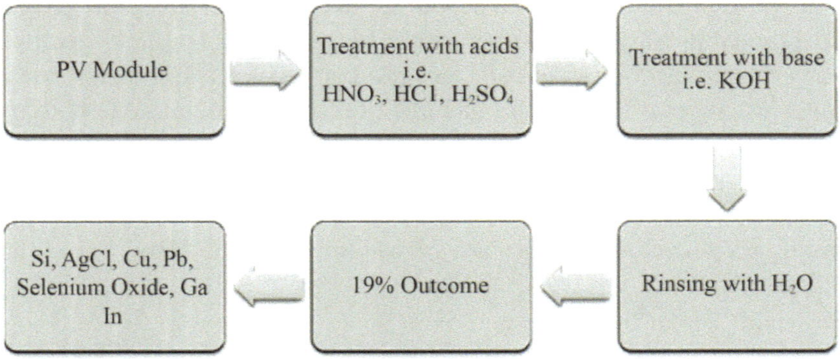

FIGURE 7.5 Chemical recycling method.

Solar module waste has detrimental effects on the environment, humans, ecosystems, and biodiversity. Major aftereffects on humans included the carcinogenic effects of Cd, Te, Se, effects of germane, indium compounds on the gastrointestinal tract, etc., [18, 19]. Figure 7.6 shows the current market share of the various solar PV technologies (2014–2030):

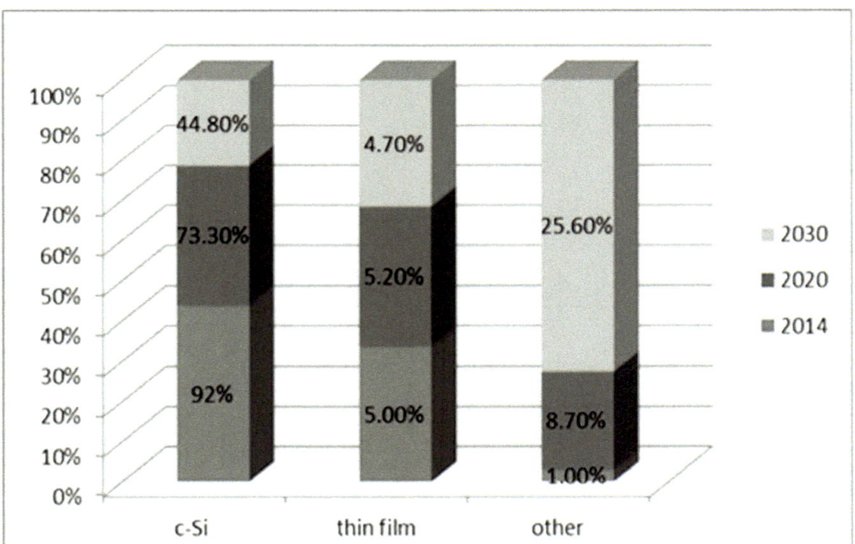

FIGURE 7.6 Market share of PV panels (2014–2030).
Source: IRENA 2016 [20].

With increasing PV installations the amount of PV waste proliferation is increasing and the recycling process is still at the nascent stage [20]. The completion of the life cycle of the solar photovoltaic module after 25–30 years has a scope of recycling but the estimation of the total outcome of recycling is a complication. Thermal and chemical-based end life treatment methods are used to curb this complication [21, 22]. Further, the total material input is analyzed and then compared with the output after recycling. The outcome ratio of elements like aluminum, silver, copper are assessed and other materials like steel and glass are also assessed [23–25].

Till the present hour, the largest fraction of waste PV materials are dumped in landfills. Though recycling could result in less environmental burden than landfilling but at an added monetary cost. So the need of the hour is to build up an effective PV waste management [26].

7.3 SOLAR CELL

A solar cell is a photovoltaic device that converts the light energy into electrical energy based on the principles of the photovoltaic effect. The generation of voltage across the PN junction in a semiconductor due to the absorption of light radiation is called a photovoltaic effect. The device based on this effect is called a photovoltaic device. The working of a solar cell is shown in Figure 7.7. The movement of the electrons takes place from the n region to the p region across the depletion region and an electric field is generated. The n and p regions are further connected to the finger electrodes to collect the electrons and are further supplied to the load.

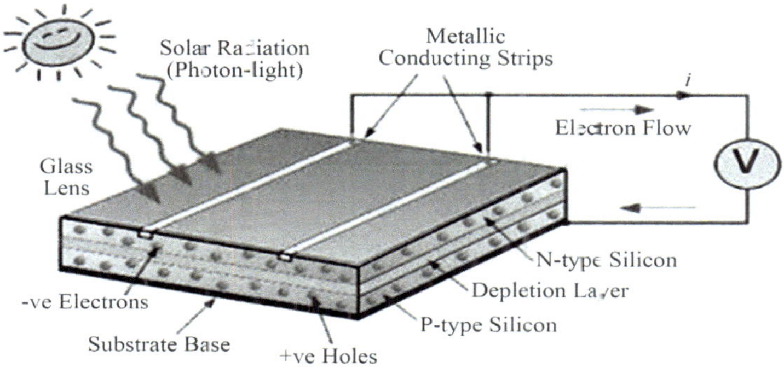

FIGURE 7.7 Working of solar cell.

The categorization of the solar photovoltaic solar cells is done on the basis of the type of materials used and is discussed below [9, 27].

7.3.1 FIRST GENERATION SOLAR CELL-WAFER BASED

These types of solar cells are made by using wafers of silicon. It's the most famous technology because of its high efficiency and is the most primitive. It is further divided into two types:

1. **Single-Crystalline Silicon Solar Cell:** Single crystal of silicon is used in making this type of solar cell and their efficiency is of the order of 17–18%.
2. **Multi-Crystalline Silicon Solar Cell:** Different crystals are combined together to make a cell. In the present time, it's the most popular. These are also cheap to compile in comparison to single crystal solar cells and their efficiency is less to nearby 12–14% [12, 16].

7.3.2 THIN FILM-BASED SOLAR CELLS

In comparison to wafer-based silicon cells, thin-film solar cells have more economical aspects as 350 micro-meter layers are present for absorbing the incident photons. Whereas only a one-micrometer thick layer is present in the wafer-based cell.

Solar cells made of thin-film are further divided into:

- Amorphous Silicon (a-Si);
- Cadmium telluride (CdTe);
- Copper indium gallium di-selenide (CIGS).

1. **Amorphous Silicon Thin Film (a-Si) Solar Cell:** The atoms of silicon are arranged in a haphazard manner in the lattice of the crystal. It is the most elementary solar cell which was firstly prepared at an industrial scale. It is relatively having a nominal cost and extensively available. Its efficiency is low from 4% to 8%.
2. **Cadmium Telluride (CdTe) Thin-Film Solar Cell:** From the rest of the thin-film solar cells cadmium telluride solar cells are

the most prominent type. These are feasible in terms of cost also. Their cost is much more reasonable in comparison to other solar cells. Their working efficiency is from 9% to 11%.

In spite of the above favorable reasons to utilize more CdTe solar cells, the transition element Cadmium present in it imparts a detrimental effect on the living beings of the ecosystem that is, plants, animals, humans, etc. The reclaiming process of Cadmium is also a cost demanding and the output ratio is also not much feasible. Transition elements also have low availability in nature. The threatening effect of Cadmium on the surroundings is also an issue of concern.

3. **Copper Indium Gallium Di-Selenide (CIGS) Solar Cell:** Such type of solar cell contains a combination of tetra elements comprising of representative elements and transition elements. These solar cells possess greater efficiency of 10% to 12% in comparison to CdTe solar cells. In terms of cost, also these solar cells are relatively cheap in comparison to their other thin-film solar cell counterparts [28, 29].

7.3.3 THIRD GENERATION SOLAR CELLS

This generation of solar cells is the emerging one in the latest hour in both the industry level and research level. Various types included in this category are:

1. **Polymer-Based Solar Cell (PSC):** This type of solar cell is made up of a thin layer of polymer wrapped on the absorbing layer.
2. **Concentrated Solar Cells (CSC):** This is the modern photovoltaic technology. In this technology, the incident energy of the photons is converted into a small area.
3. **Dye-Sensitized Solar Cells (DSC):** This technology utilizes dye molecules in between the collecting electrodes.
4. **Solar Cells Based on Perovskite:** In this type of solar cell halogen ions like iodide, bromide, chloride, and two different cations, that is, A and B of variable radius are present. Usually, ABX_3 formula compounds are included. This perovskite solar technology is the recent advancement in the photovoltaic industry.

5. **Nano Crystal-Based Solar Cells:** This category of the solar cell consists of semiconductor material in the nanometer range. Elements of the transition group are also present. Solar cells based on nanocrystals are also known as quantum dot cells.

All these above mentioned solar photovoltaic panels contain the various elements as mentioned below in Table 7.1 along with their production and their reserves in million tons from worldwide as documented by the U.S. Department of the Interior and U.S. Geological Survey in the Mineral Commodity Summaries-2019.

The first shipment of the PV installations from the early 1980 has already withdrawn from their active life. And there is an urgency to handle the PV waste generated. The waste so generated at the end is of two types:

1. Manufacturing waste generated in the plant; and
2. Waste generated at the end of life.

TABLE 7.1 Production and Reserve of Elements Used in PV Panels

	Mine Production		
Element	**2017**	**2018**	**Reserves**
Copper	20,000	21,000	830,000
Aluminum	59,000	60,000	80,300
Arsenic	34,600	35,000	N.A
Cadmium	25,400	26,000	N.A
Gallium	320,000	410,000	N.A
Germanium	106,000	120,000	N.A
Indium	714	750	N.A
Selenium	2710	2800	99,000
Silver	26,800	27,000	560,000
Silicon	6580	6700	N.A

By 2030, 1.7 million tons, and by 2050, 60 million tons of PV waste will be generated. Analytical studies put forward on thin-film solar cells showed that 97% of the materials used in their manufacturing can be removed from the whole [26, 30, 31]. However, solar energy is taken as the most reassuring technology for cleaner energy production. Further, the performance of the PV technology is assessed in terms of environmental conduct.

7.4 ENVIRONMENTAL REPERCUSSION

The environmental impact of the c-Si and thin-film PV modules is analyzed through the life cycle assessment. GaBi is used for the LCA model assessment, CML 2001 baseline method is utilized for the assessment of the environmental effects of the modules. CdTe modules though require less material input than c-Si technology but the catastrophic aftermath of the CdTe modules is much greater than that laid out by the c-Si [32].

Using transition metals like cadmium, copper, and other elements like tellurium, selenium, silicon in the manufacturing of bulk and thin-film solar photovoltaic materials causes major concern for health and environment at their end of life phase [22]. During the operation phase, bulk and thin-film solar photovoltaic panels cause no harm to the environment. Exposure to arsine, cadmium, germane, lead, phosphorous oxychloride causes issues related to kidney which can further lead to nephrotoxicity [15, 33, 34]. Arsine and carbon tetrachloride can cause a severe effect on the lungs. Hydrogen fluoride and compounds of indium used in the manufacturing of thin-film photovoltaic materials cause detrimental effects on bones and teeth. Deposition of fluoride and indium compounds on bones and teeth causes skeletal and dental fluorosis. Exposure to phosphine causes cardiovascular dysfunction, gastrointestinal disorder; it can also act as a pulmonary irritant. Phosphine can also catch fire due to a sudden rise in the ambient temperature, which can lead to hemorrhage, neuropsychiatric disorders, respiratory and renal failures within a few hours of exposure [35]. Inhalation of germane during the manufacturing of thin-film photovoltaic materials can lead to dizziness, abdominal pain, and headache. Germane catches fire quickly on exposure to air; it can also lead to an explosion on exposure to high temperatures and can also cause a hazard. Long term exposure to germane and arsine causes lesions of blood cells which result in decreased efficiency to carry blood. Indium compounds can also cause irritation to the eyes, skin, and esophagus. Silane, selenium oxides, and selenium hydroxides on inhalation can cause irritation to the skin, eyes, and mucous membrane [22, 36]. The major emphasis in this section is given on human health during the manufacturing and after the recycling process. The environmental aftereffects on humans included the carcinogenic effects of Cd, Te, Se, arsine, carbon tetrachloride, etc., and other health issues as discussed above on various body organs are shown in Figure 7.8.

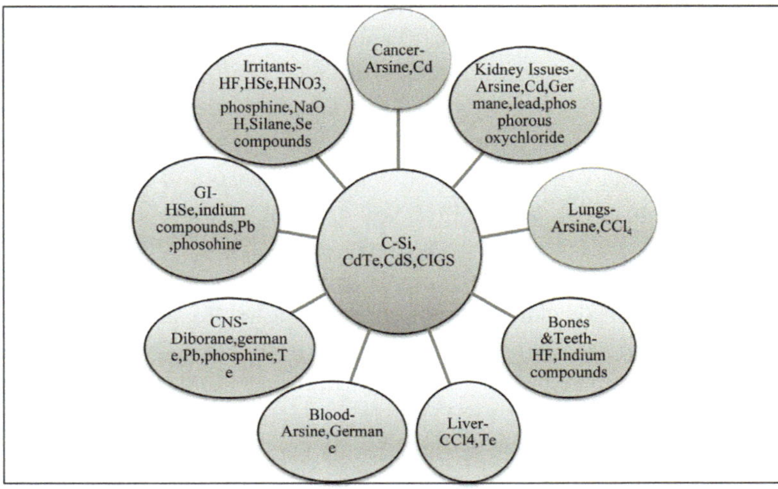

FIGURE 7.8 Effects on health by solar PV modules.

7.5 LEGISLATIVE FRAMEWORK FOR SOLAR PV RECYCLING IN INDIA

With the macroscopic proliferation of photovoltaic solar panel waste (PV material waste) in our country, it has become an utmost need of the hour to establish a PV scrap management and recycling policy. In India, the solar photovoltaic material recycling is still at its infancy [37]. In the present time, India is not having any lawmaking for the compulsory assemblage, resumption, and recycling of life span of PV material. Bridge to India managing committee aforementioned that while the solar sector lengthens to heighten greatly from bare 3 gigawatts in 2014 to around 28 gigawatts in 2019.

Currently, in India, there is a lack of transparency in solar waste management. The endowment of the sustainable energy pillar act is required so that the PV scarp is extracted out from the general waste systemization. Solar PV material recycling is still not with peculiar laws

[38]. WEEE published a Directive 2002/19/EC in the official journal of the European Union on 4 July 2012. The above directive was retrieved from the 2002/96/EC directive; this directive includes the photo-voltaic panels into the electrical and electronic equipment catalog. For the past 10 years, the European Union policy accords emphasis on recycling and resumption of the material and least on landfilling Under the 2016 solid waste management rules and transboundary movement rules, the ministry of environment forest, and climate change is managing the other waste and hazardous waste. Solar PV material manufacturing utilizes hazardous compounds, if these are left without treatment or recycling would cause adverse environmental impact [39].

At present, India neither possesses a policy guideline nor the least operational groundwork to safeguard the recycling of PV scrap by the traditional recycling process. With the sky rocking PV scrap, India is obligate for inclination towards solar PV recycling policy framework. Solar PV recycling legislative recommendations can be urged from members across the globe. Below are the ideas for the Indian government to tackle the inevitable PV scarp dilemma by applying the recycling legislative strategy in a detailed manner:

1. **Regulative Framework:**
   - The manufacturing firm should be registered with a license.
   - Constituent components utilization should also be familiarized.
   - Prohibit the use of low abundant elements.
   - Warranty of accomplishing environmentally loyal materials
   - Provision of end life sketch for retiring panels.
   - Comprehensively manufacturing should be amiable to nature.
   - Provision of analytical testimony from birth to retirement.
   - Gross production & import of all PV material trading routes should be monitored.
2. **Layout the Accountability and Obligation of the Collaborator for Scrap Handling and Operation:**
   - Establishment of joint collection ventures across India for the retired and distorted solar panels between the producer and the recycler [40].
   - Consumers should be provided some incentives for waste deposition in collection centers.

- Collected waste material can then be auctioned for the recyclers or tenders can be directed.
- Collection centers should have familiarity with waste disposal regulations [20].
- Alimony to such joint ventures should be provided.

3. **Compose Model for PV Waste Compilation:**
   - B2B (Business to Business) and B2C (Business to consumer) firms should be utilized.
   - A municipal collection point is utilized.
4. **Launch of Treatment Method:**
   - A combination of both mechanical and thermal methods should be registered commercially.
   - Method should deal with hazardous and non-hazardous waste.
   - Risk factors of treatment of the wastes should be established to the recyclers.
   - Standardized infrastructure for the treatment should be available.
5. **Disposition Corporation Endowment:**
   - Survey the usual recycling methods.
   - Analyze the expenditure and the industrial need for the efficient PV recycling system.
   - Develop units to accumulate recycling materials.
   - Supply recycled products to related industries for further profit gain.
6. **Manufacturer Liability System (MLS) Constitution:**
   - Abide the manufacturer to ensure a lesser impact of hazardous materials.
   - Utmost stringent responsibility of the producer to take back its product from the market after its complete life end.
7. **Set up of a Head-Body Overall the Solar PV Industry:**
   - Monitor the overall activity of the solar firms and knock out offenders from registration.
   - This organization should also promote the use of sustainable materials in the design.

8. **Financial Aid:**
   - A federal government agency should be set up for the financial assessment of the PV industry.
   - Indian government should inculcate various monetary schemes for recyclers.
   - These ways hazardous effects of the clean energy source are totally culminated out from the environment.
9. **Recycling Institution Set-Up:**
   - India should establish institutes which will impart professional skills among the people involved in this field.
   - It will also be a boon for employment.

7.6 CONCLUSION

The chapter presents a policy framework for solar PV recycling in both rural and urban India. With everyday target achievement of new solar PV plants, the waste accumulation has reached to an estimated amount of 6,096,508 tons. Till now, the recycling methods have received very little emphasis. Treacherous planetary waste spawned by the decommissioning of solar PV materials can't be thrown in the garbage. There has to be a stringent plan for its management. The implementation of a legislative PV recycling structure in emergent nations especially in the Indian scenario will be a great life savior to tackle the havoc which is to be originated in times to come. A comprehensive analysis is carried out on existing PV recycling methods out of them thermal method is good enough with some demerits. The scrutiny of LCA methodology of PV comprehends the three phases: (1) manufacturing, (2) utility, and (3) end life.

Manufacturing phase is more influential for the environmental perspective, as it holds the involvement of various substances in all states of matter which can produce endangerment to the health of humans. The utility phase can also add to the harmful emissions due to fire hazards which can further cause harm to the land. The end life phase accumulates the waste pile up. This chapter also emphasizes on the environmental and socio-economic tremors bumped by the clean energy source due to the increasing market share of PV technology. An affirmative response to this policy from the Indian government will effectively curb the future forecast of the PV scrap disaster. Harmful gaseous emissions are released into the environment during the process of manufacturing, utility phase, and end life phase. Further, not only the manufacturing process but also the resource consumption is also considered and the impact is evaluated in context to the environment. A complete interpretation is made for the disposal and end life plot of c-Si and CdTe panel. Recycling of the solar photovoltaic modules at the end of life plays an important role in the restoration of the materials used. Recycling is veiled up by the complications encountered during the procedure like lack of collection point of the waste and proper recycling technology. Inadequacy of the material availability is paving the way towards the recycling of the material from the PV modules. Recycling of the solar panels can help in abstaining the material wastage.

KEYWORDS

- **environmental impact**
- **legislative framework**
- **life cycle assessment**
- **PV recycling**
- **selenium hydroxides**
- **solar energy**
- **solar e-waste**

REFERENCES

1. Fthenakis, V. M., & Kim, H. C., (2011). Photovoltaics: Life-cycle analyses. *Solar Energy*, *85*(8), 1609–1628. https://doi.org/10.1016/j.solener.2009.10.002 (accessed on 18 July 2020).
2. Baharwani, V., Meena, N., Dubey, A., Sharma, D., Brighu, U., & Mathur, J., (2014). Life cycle inventory and assessment of different solar photovoltaic systems. In: *2014 Power and Energy Systems: Towards Sustainable Energy* (pp. 1–5). Bangalore, India: IEEE. https://doi.org/10.1109/PESTSE.2014.6805302 (accessed on 18 July 2020).
3. Palitzsch, W., & Loser, U., (2012). Economic PV waste recycling solutions-results from R&D and practice. In: *2012 38th IEEE Photovoltaic Specialists Conference* (pp. 628–631). Austin, TX, USA: IEEE. https://doi.org/10.1109/PVSC.2012.6317689 (accessed on 18 July 2020).

4. Fiandra, V., Sannino, L., Andreozzi, C., & Graditi, G., (2019). End-of-life of silicon PV panels: A sustainable materials recovery process. *Waste Management*, *84*, 91–101. https://doi.org/10.1016/j.wasman.2018.11.035 (accessed on 18 July 2020).
5. Vellini, M., Gambini, M., & Prattella, V., (2017). Environmental impacts of PV technology throughout the life cycle: Importance of the end-of-life management for Si-panels and CdTe-panels. *Energy*, *138*, 1099–1111. https://doi.org/10.1016/j.energy.2017.07.031 (accessed on 18 July 2020).
6. Luo, W., Khoo, Y. S., Kumar, A., Low, J. S. C., Li, Y., Tan, Y. S., Wang, Y., Aberle, A. G., & Ramakrishna, S., (2018). A comparative life-cycle assessment of photovoltaic electricity generation in Singapore by multi-crystalline silicon technologies. *Solar Energy Materials and Solar Cells*, *174*, 157–162. https://doi.org/10.1016/j.solmat.2017.08.040 (accessed on 18 July 2020).
7. Vargas, C., & Chesney, M., (2019). End of life decommissioning and recycling of solar panels in the United States. A real options analysis. *SSRN Journal*. https://doi.org/10.2139/ssrn.3318117 (accessed on 18 July 2020).
8. Frischknecht, R., Itten, R., Sinha, P., De Wild-Scholten, M., Zhang, J., Heath, G. A., & Olson, C., (2015). *Life Cycle Inventories and Life Cycle Assessments of Photovoltaic Systems*. Report No. NREL/TP-6A20–73853, 1561526. https://doi.org/10.2172/1561526(accessed on 18 July 2020).
9. Lamnatou, C., & Chemisana, D., (2019). Life-cycle assessment of photovoltaic systems. In: *Nanomaterials for Solar Cell Applications* (pp. 35–73). Elsevier. https://doi.org/10.1016/B978-0-12-813337-8.00002-3 (accessed on 18 July 2020).
10. Deng, R., Chang, N. L., Ouyang, Z., & Chong, C. M., (2019). A techno-economic review of silicon photovoltaic module recycling. *Renewable and Sustainable Energy Reviews*, *109*, 532–550. https://doi.org/10.1016/j.rser.2019.04.020 (accessed on 18 July 2020).
11. Bilich, A., Goyal, L., Hansen, J., Krishnan, A., & Langham, K., (2016). *Assessing the Life Cycle Environmental Impacts and Benefits of PV-Micro Grid Systems in Off-Grid Communities*, *160*.
12. Tammaro, M., Salluzzo, A., Rimauro, J., Schiavo, S. & Manzo, S., (2016). Experimental investigation to evaluate the potential environmental hazards of photovoltaic panels. *Journal of Hazardous Materials*, *306*, 395–405. https://doi.org/10.1016/j.jhazmat.2015.12.018 (accessed on 18 July 2020).
13. Latunussa, C. E. L., Ardente, F., Blengini, G. A., & Mancini, L., (2016). Life cycle assessment of an innovative recycling process for crystalline silicon photovoltaic panels. *Solar Energy Materials and Solar Cells*, *155*, 101–111. https://doi.org/10.1016/j.solmat.2016.03.020 (accessed on 18 July 2020).
14. Klugmann-Radziemska, E., & Ostrowski, P., (2010). Chemical treatment of crystalline silicon solar cells as a method of recovering pure silicon from photovoltaic modules. *Renewable Energy*, *35*(8), 1751–1759. https://doi.org/10.1016/j.renene.2009.11.031 (accessed on 18 July 2020).
15. Huang, B., Zhao, J., Chai, J., Xue, B., Zhao, F., & Wang, X., (2017). Environmental influence assessment of China's multi-crystalline silicon (Multi-Si) photovoltaic modules considering the recycling process. *Solar Energy*, *143*, 132–141. https://doi.org/10.1016/j.solener.2016.12.038.

16. Phylipsen, G. J. M., & Alsema, E. A., (1995). *Environmental Life-Cycle Assessment of Multi Crystalline Silicon Solar Cell Modules, 66.*
17. Huang, B., Zhao, J., Chai, J., Xue, B., Zhao, F., & Wang, X., (2017). Environmental influence assessment of china's multi-crystalline silicon (Multi-Si) photovoltaic modules considering the recycling process. *Solar Energy, 143*, 132–141. https://doi.org/10.1016/j.solener.2016.12.038 (accessed on 18 July 2020).
18. Goe, M., & Gaustad, G., (2016). Estimating direct human health impacts of end-of-life solar recovery. *Golisano Institute for Sustainability.*
19. Gottesfeld, P., & Cherry, C. R., (2011). Lead emissions from solar photovoltaic energy systems in China and India. *Energy Policy, 39*(9), 4939–4946. https://doi.org/10.1016/j.enpol.2011.06.021 (accessed on 18 July 2020).
20. Weckend, S., Wade, A., & Heath, G. A., (2016). *End of Life Management: Solar Photovoltaic Panels.* Report No. NREL/TP-6A20–73852, 1561525. https://doi.org/10.2172/1561525 (accessed on 18 July 2020).
21. Klugmann-Radziemska, E., (2012). Current trends in the recycling of photovoltaic solar cells and modules waste/recyklingzużytychogniwimodułówfotowoltaicznych-stan obecny. *Chemistry-Didactics-Ecology-Metrology, 17*(1/2), 89–95. https://doi.org/10.2478/cdem-2013–0008 (accessed on 18 July 2020).
22. Quek, T. Y. A., Alvin, E. W. L., Chen, W., & Ng, T. S. A., (2019). Environmental impacts of transitioning to renewable electricity for Singapore and the surrounding region: A life cycle assessment. *Journal of Cleaner Production, 214*, 1–11. https://doi.org/10.1016/j.jclepro.2018.12.263 (accessed on 18 July 2020).
23. Islam, M. T., & Huda, N., (2019). E-waste in Australia: Generation estimation and untapped material recovery and revenue potential. *Journal of Cleaner Production, 237*, 117787. https://doi.org/10.1016/j.jclepro.2019.117787 (accessed on 18 July 2020).
24. Larsen, K., (2009). End-of-life PV: Then what? *Renewable Energy Focus, 10*(4), 48–53. https://doi.org/10.1016/S1755–0084(09)70154–1 (accessed on 18 July 2020).
25. Chen, W., Hong, J., Yuan, X., & Liu, J., (2016). Environmental impact assessment of monocrystalline silicon solar photovoltaic cell production: A case study in China. *Journal of Cleaner Production, 112*, 1025–1032. https://doi.org/10.1016/j.jclepro.2015.08.024 (accessed on 18 July 2020).
26. Anctil, A., & Fthenakis, V., (2013). Critical metals in strategic photovoltaic technologies: Abundance versus recyclability: Critical metals in strategic photovoltaic technologies. *Prog. Photovolt: Res. Appl., 21*(6), 1253–1259. https://doi.org/10.1002/pip.2308 (accessed on 18 July 2020).
27. Paiano, A., (2015). Photovoltaic waste assessment in Italy. *Renewable and Sustainable Energy Reviews, 41*, 99–112. https://doi.org/10.1016/j.rser.2014.07.208 (accessed on 18 July 2020).
28. Azeumo, M. F., Germana, C., Ippolito, N. M., Franco, M., Luigi, P., & Settimio, S., (2019). Photovoltaic module recycling, a physical and a chemical recovery process. *Solar Energy Materials and Solar Cells, 193*, 314–319. https://doi.org/10.1016/j.solmat.2019.01.035.
29. Meinert, L., Robinson, G., & Nassar, N., (2016). Mineral resources: Reserves, peak production, and the future. *Resources, 5*(1), 14. https://doi.org/10.3390/resources5010014 (accessed on 18 July 2020).

30. Peeters, J. R., Altamirano, D., Dewulf, W., & Duflou, J. R., (2017). Forecasting the composition of emerging waste streams with sensitivity analysis: A Case study for photovoltaic (PV) panels in flanders. *Resources, Conservation and Recycling*, *120*, 14–26. https://doi.org/10.1016/j.resconrec.2017.01.001 (accessed on 18 July 2020).
31. Ilias, A. V., Meletios, R. G., Yiannis, K. A., & Nikolaos, B., (2018). Integration and assessment of recycling into the C-Si photovoltaic module's life cycle. *International Journal of Sustainable Engineering*, *11*(3), 186–195. https://doi.org/10.1080/19397038.2018.1428833 (accessed on 18 July 2020).
32. Weber, R. J., & Reisman, D. J., (2012). *Rare Earth Elements: A Review of Production Processing, Recycling, and Associated Environmental Issues*, *21*.
33. Savvilotidou, V., Antoniou, A., & Gidarakos, E., (2017). Toxicity assessment and feasible recycling process for amorphous silicon and CIS waste photovoltaic panels. *Waste Management*, *59*, 394–402. https://doi.org/10.1016/j.wasman.2016.10.003 (accessed on 18 July 2020).
34. Faircloth, C. C., Wagner, K. H., Woodward, K. E., Rakkwamsuk, P., & Gheewala, S. H., (2019). The environmental and economic impacts of photovoltaic waste management in Thailand. *Resources, Conservation and Recycling*, *143*, 260–272. https://doi.org/10.1016/j.resconrec.2019.01.008 (accessed on 18 July 2020).
35. Lee, S. H., & Kang, H. G., (2014). *Comparative Health Risk Assessment of CdTe Solar PV System and Nuclear Power Plant*, *10*.
36. Faircloth, C. C., Wagner, K. H., Woodward, K. E., Rakkwamsuk, P., & Gheewala, S. H., (2019). The environmental and economic impacts of photovoltaic waste management in Thailand. *Resources, Conservation and Recycling*, *143*, 260–272. https://doi.org/10.1016/j.resconrec.2019.01.008 (accessed on 18 July 2020).
37. IEA, P., (2019). *Snapshot of Global PV Market*. Strategy PV analysis and outreach; snapshot T1:35; International Energy Agency: Spain.
38. Karthikeyan, L., Suresh, V., Krishnan, V., Tudor, T., & Varshini, V., (2018). The management of hazardous solid waste in India: An Overview. *Environments*, *5*(9), 103. https://doi.org/10.3390/environments5090103 (accessed on 18 July 2020).
39. Tasnia, K., Begum, S., Tasnim, Z., & Khan, M. Z. R., (2018). End-of-life management of photovoltaic modules in Bangladesh. In: *2018 10th International Conference on Electrical and Computer Engineering (ICECE)* (pp. 445–448.). Dhaka, Bangladesh: IEEE. https://doi.org/10.1109/ICECE.2018.8636782. (accessed on 18 July 2020).
40. Xu, Y., Li, J., Tan, Q., Peters, A. L., & Yang, C., (2018). Global status of recycling waste solar panels: A review. *Waste Management*, *75*, 450–458. https://doi.org/10.1016/j.wasman.2018.01.036 (accessed on 18 July 2020).

CHAPTER 8

Simulation of Multilayer Atom Nanostructures for Spinmechatronics

A. V. VAKHRUSHEV,[1,2,3] A. YU. FEDOTOV,[1,2,3] V. I. BOIAN,[4] and A. S. SIDORENKO[3,4,5]

[1]*Department of Modeling and Design of Technological Structures, Institute of Mechanics, Udmurt Federal Research Center, Ural Division, Russian Academy of Sciences, Izhevsk, Russia, E-mail: Vakhrushev-a@yandex.ru*

[2]*Department of Nanotechnology and Microsystems, Kalashnikov Izhevsk State Technical University, Izhevsk, Russia*

[3]*Department of Functional Nanostructures, Orel State University named after I.S. Turgenev, Orel, Russia*

[4]*Department of Cryogenics, Ghitu Institute of Electronic Engineering and Nanotechnologies, Chisinau, Republic of Moldova*

[5]*Department of Microelectronics and Biomedical Engineering, Technical University of Moldova, Chisinau, Republic of Moldova*

ABSTRACT

The work is devoted to the study of the formation processes and analysis of the structure of a superconducting spin valve based on a multilayer superconductor-ferromagnet nanostructure. The relevance of the research is due to the need to develop an energy-efficient element base for microelectronics, based on new physical principles and the advent of devices, based on spin and quantum-mechanical effects. The superconducting spin valve being developed is a multilayer structure consisting of ferromagnetic cobalt nanofilms, which are separated by niobium superconductors. The studies were carried out using molecular dynamics modeling. As the interaction potential of atoms in the simulated system, the modified immersed atom method is used. The spin valve was formed by layer-by-layer deposition of elements in a vacuum. The atom deposition process was simulated in a stationary temperature regime. The chapter presents a simulation of the deposition of the first few layers of a nanosystem. The atomic structure of individual nanolayers of the system is considered. Particular attention is paid to the analysis of the atomic structure of contact areas at the junction of the layers since the quality of the layer interface plays a crucial role in creating a workable device. Three temperature deposition regimes were implemented: 300, 500, and 800 K. Calculations showed that with an increase in temperature, there is a rearrangement of the structure of the system layers and their loosening. The structure of the nanolayer from niobium is close to crystalline with division into regions of different crystallographic orientations of atomic layers. For cobalt nanofilms, an amorphous structure is more characteristic. The obtained simulation results can be used in development. as well as optimization of technologies for the formation of spin valves and other functional elements for spintronics.

8.1 INTRODUCTION

With the development of nanotechnology, a significant interest has appeared in science and industry in a special section of quantum electronics, which studies the transfer of spin current in solids and is called spintronics [1–3]. Unlike conventional electronics, the target of research in spintronics is not an electric current, but a spin current, which can be used to transport information. The basis of systems in spintronics is heterostructures consisting of ferromagnets, superconductors, and normal metals (paramagnets) [4, 5], which suggests the creation of multilayer nanocomposites formed by nanofilms with quite different physical properties.

Currently, a prototype of a nanoscale battery with a significant electromotive force of a spin nature is known [6, 7]. The complexity of the production of spin batteries is due to the need to form a special magnetic tunnel junction, which is a design feature of such devices. The presence of a tunnel junction strongly depends on the starting materials of the heterostructures and their properties.

The use of spin effects for storing and processing information is promising. The authors of Refs. [8, 9] showed that the hexagonal structure of nanoscale magnets under the external influence can be rebuilt with the formation of phase transitions, fluxes of magnetic particles, and magnetic defects. Changes in the states of individual magnetic cells occur along the chain and affect the polarity of neighboring regions, which resembles the principle of signal transmission by neurons and underlies the storage and processing of data. The operating temperature of the device is significantly lower than room temperature; therefore, the authors of the development called the created magnetic processor "artificial spin ice."

Devices based on manipulating the spins of particles consume incomparably less energy than traditional semiconductor microelectronic devices, and, therefore, these materials serve as a promising alternative to elements from conventional electronics, generating a potential difference due to spin-polarized transport. As described in Ref. [10], the source of spin-polarized electrons can be a magnetized rod of a nickel-iron alloy. Heating and cooling of the opposite ends of the rod leads to the redistribution of electrons with different spins and, therefore, creates an electric voltage. In the future, on the basis of such schemes, it is possible to build powerful and small-sized computers.

Over the past decade, a new direction of research has been formed [11, 12], in which materials under the influence of external electric and magnetic fields are able to change the electrical resistance by several orders of magnitude. This phenomenon has received the name of colossal magnetic resistance and has significant prospects for creating new technologies for recording and storing information with increased density and low power consumption compared to conventional modern data carriers. The study of a crystalline sample of manganite of perovskite structure by the method of built-in combination of a scanning tunneling microscope and an electron microscope revealed a distortion of the crystal lattice of a substance caused by the combined motion of electrons and phonons. Materials with colossal magnetic resistance can be used to create computer non-volatile memory (MRAM), as well as in other spintronics and electronics devices.

Despite the variety of spin, effects and devices realized with their use, one of the fundamental values is the structure of the magnetic material. As noted in Refs. [13–15], at the interface between a superconductor and a ferromagnet, a region of mutual influence of competing states arises, which, under the action of a directed magnetic field, makes it possible to switch the heterostructure from the normal state to the superconducting and vice versa [16–19]. The study of the structure of the intermediate region is the subject of intensive theoretical [13] and experimental [14,

20] studies. In studies [14], it was demonstrated that the interface between a superconductor and a ferromagnet affects not only the characteristics of the contact layer but also leads to a change in the properties of the vicinity of this adjacent region. In addition, in real materials, the boundary of the contact layer is quite extended and structurally inhomogeneous. The magnetic properties of the final material are directly dependent on nanometer magnetic clusters and ferromagnetic domains in the contact zone, as well as their interaction with each other [21].

In connection with the foregoing, the urgent task of a detailed study of the structure contact layer of layered materials superconductor-ferromagnet and its definition spatial profile. To build methodological foundations and conduct theoretical studies of the structure of multilayer spin systems in this paper simulated nanosystem with a controlled effective energy of exchange for devices memory and switches type "spin valve" [22, 23], which is a multilayer structure consisting of ferromagnetic cobalt nanofilms that are separated by thin layers of niobium superconductor.

The aim of this study is the simulation processes of the formation of multilayer atom nanostructures for spintronics. The objects of the investigation were the structure of a superconducting spin valve based on a multilayer superconductor-ferromagnet nanostructure and interface between the layers of the spin vale.

This work is a continuation and development of research on modeling of formation processes, atomic structure, and properties of various complex multiphase nanosystems [24], prepared in a vacuum and in the gas phase [25, 26], in halocarbon and metal-organic systems [27–29], on continuous [30, 31] and porous substrates [32].

8.2 MATHEMATICAL MODEL AND STATEMENT OF THE PROBLEM

General scheme of a magnetic heterostructure indicating magnetic field strength shown in Figure 8.1. In this work, calculations were carried out at zero magnetic field H = 0. The numbers next to the elements in the layers show their thickness in nanometers. Sample preparation takes place by material deposition under conditions close to the vacuum. In the general case, the nanosystem contains more than 20 layers, but their processes formations, as well as their structural features, are similar. Therefore, in this work deposition of only the first few layers of the sample will be considered.

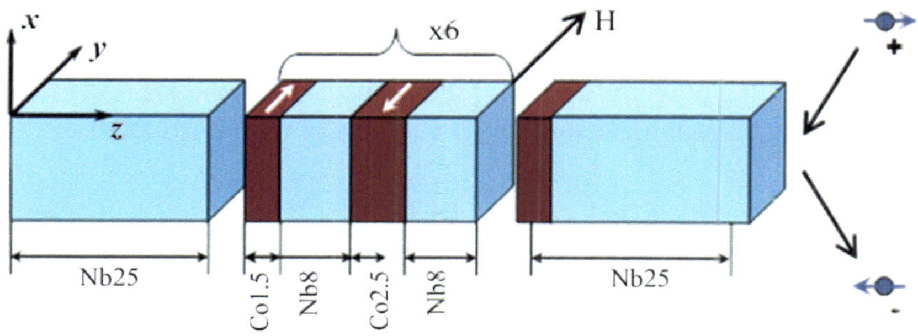

FIGURE 8.1 The structural scheme of a multilayer magnetic composite is based on cobalt and niobium used to create memory devices.

The contact layer between superconducting and ferromagnetic materials was studied by the molecular dynamics method. In this case, both the formation process of multilayer nanofilms and the resulting structure formed by atoms inside a multilayer nanocomposite were considered. The molecular dynamics method is based on the equation of Newton's motion, which is solved for each atom:

$$m_i \frac{d^2\mathbf{r}}{dt^2} = -\frac{\partial U(\mathbf{r})}{\partial \mathbf{r}_i} + \mathbf{F}_{ex}, \quad \mathbf{r}_i(t_0) = \mathbf{r}_{i0}, \quad \frac{d\mathbf{r}_i(t_0)}{dt} = \mathbf{V}_{i0}, \quad i = 1, ..N \tag{1}$$

where, N is the number of atoms that formed nanosystem; m_i is the mass of the i-th atom; $\mathbf{r}_{i0}, \mathbf{r}_i(t)$ – are the initial and current radius vector of the i-th atom, respectively; $U(\mathbf{r}(t))$ is the potential energy of the system; $\mathbf{V}_{i0}, \mathbf{V}_i(t)$ are the initial and current speed of the i-th atom, respectively; $\mathbf{r}(t) = \{\mathbf{r}_1(t), \mathbf{r}_2(t), .., \mathbf{r}_N(t)\}$ shows the dependence of the location of all the atoms system; and $\mathbf{F}(\mathbf{r}(t), t)$ is an external force.

For definiteness of the solution of the molecular dynamics equation, it is necessary to have specifying conditions, which in Eqn. (1) are the indication of the initial coordinates and velocities for all atoms.

The results of solving the problems of molecular dynamics strongly depend on the type and accuracy of the potential field U(r) in Eqn. (1). Currently, there are a large number of varieties of potentials, both paired and multiparticle. A well-established potential with a wide range of applications is a modified embedded atom method (MEAM). This potential dynamically simulates the formation and breaking of bonds between atoms, considers the asymmetric directivity of electron clouds, and is able to adequately reproduce the physical properties of complex crystalline metals and semiconductors [33, 34].

The resulting potential energy of the nanosystem in the MEAM method is represented through the sum of the energies of individual atoms.

$$U(r)=\sum_i(r)=\sum_i\left(F_i(\bar{\rho}_i)+\frac{1}{2}\sum_{j^1i}f_{ij}(r_{ij})\right), i=1,2,..., N, \quad (2)$$

where, $U_i(r)$ is the potential of the i-th atom affects the type of interaction of atoms and the magnitude of the forces in the equations of motion; F_i is the immersion function of the i-th atom located at a point in space with electron background density $\bar{\rho}_i$; $\phi_{ij}(r_{ij})$ is the value of the pair potential between the i-th and j-th atoms remote at a distance r_{ij}.

The immersion function depends on the background electron density, has a variable form for different types of chemical elements of the periodic system, and is written using the expression

$$F_i(\bar{\rho}_i)=\begin{cases} A_iE_i^0(\bar{\rho}_i)\ln(\bar{\rho}_i), \bar{\rho}_i\geq 0 \\ -A_iE_i^0\bar{\rho}, \bar{\rho}<0 \end{cases}, \quad (3)$$

where, A_i is the empirical parameter of the potential field; E_i^0 is the value of the energy of sublimation; $\bar{\rho}_i$ is the background electron density; i is an index indicating belonging to a certain type of atom.

The background electron density at the immersion point is determined by the following functional dependence:

$$\bar{\rho}_i=\frac{\rho_i^{(0)}}{\rho_i^0}G(\Gamma_i), \Gamma_i=\sum_{k=1}^{3}t_i^{(k)}\left(\frac{\rho_i^{(k)}}{\rho_i^{(0)}}\right)^2, \quad (4)$$

where, indices $k = 1, 2, 3$ correspond to *p-, d-, f-* electronic orbitals of the *i*-th atom; $t_i^{(k)}$ is the weighting factors of the model; ρ_i^0 is a background electron density of the original structure; $\rho_i^{(k)}$ are parameters characterizing the deviation of the electron density from its ideal state when all atoms are in the nodes of the crystal lattice.

Various formulations are used to calculate the function $G(\Gamma)$. The total background electron density $\bar{\rho}_i$ contains the partial contributions of the individual densities of atomic orbitals. Atomic orbitals are divided into spherically symmetric *s*-, which the electron density $\rho_i^{(0)}$ corresponds to, and angular *p*-, *d*-, *f*- clouds, with distributions $\rho_i^{(1)}, \rho_i^{(2)}, \rho_i^{(3)}$.

To determine the weighting coefficients of the model (4) function is used:

$$t_i^{(k)} = \frac{\sum_{i \neq j} t_{0,j}^{(k)} \rho_j^{A(0)} S_{ij}}{\sum_{i \neq j} \left(t_{0,j}^{(k)}\right)^2 \rho_j^{A(0)} S_{ij}} \tag{5}$$

where, $t_{0,j}^{(k)}$ are the parameters depending on the chemical type of the *j*-th element.

Together with MEAM potential, the screening function (6) is used, which is used to reduce computational time and reduce potential error:

$$S_{ij} = f_c\left(\frac{r_c - r_{ij}}{\Delta r}\right) \prod_{k \neq i,j} f_c\left(\frac{C_{ikj} - C_{\min,\,ikj}}{C_{\max,\,ikj} - C_{\min,\,ikj}}\right), \tag{6}$$

$$C_{ikj} = 1 + 2\frac{r_{ij}^2 r_{ik}^2 + r_{ij}^2 r_{jk}^2 - r_{ij}^4}{r_{ij}^4 - \left(r_{ik}^2 - r_{jk}^2\right)^2}, \tag{7}$$

$$f_c(x) = \begin{cases} 1, \ x \geq 1 \\ \left[1 - (1-x)^4\right]^2, \ 0 < x < 1, \\ 0, \ x \leq 0 \end{cases} \tag{8}$$

where, r_c is the cutoff radius of the potential; coefficients $C_{\min}$, $C_{\max}$ are formulated for each triple of atoms *i*, *j*, *k* and depend on their chemical types; Δr determines the distance exceeding the cutoff radius at which the smoothing of the force field takes place.

The molecular dynamics equation (1) is solved numerically taking into account the initial conditions and the potential of the modified submerged atom (2)–(8). As a result of numerical studies, detailed information about all atoms of the nanosystem becomes known at each moment in time.

The main basic variables are the velocities and coordinates of atoms, as well as the forces acting between them. Based on the calculation of the obtained variables, the structure of the multilayer nanocomposite material is analyzed, its dimensional characteristics are calculated, and mechanical properties and the presence of defects are determined.

The general statement of the problem of the formation of a multilayer nano heterostructure is presented in Figure 8.2. The design diagram is rotated 90° relative to the structural diagram shown in Figure 8.1.

The first layer of material formed by niobium atoms is the substrate and the basis for the vacuum deposition of subsequent nanofilms of cobalt and niobium. The substrate is located in the lower part of the computational cell; its lowermost layer is fixed in order to exclude random movement of the sample during the modeling process.

In horizontal directions, periodic boundary conditions are imposed on the computational cell, which makes it possible to reduce the computational domain and reduce the computation time. In the upper region, boundary reflection conditions are present so that the deposited atoms do not leave the modeling system.

The deposition process is modeled by introducing atoms in the region above the substrate. In this case, the deposited atoms are given speed towards the substrate. The layers are sprayed over time in stages, due to changes in the types of atoms on the substrate cobalt or niobium.

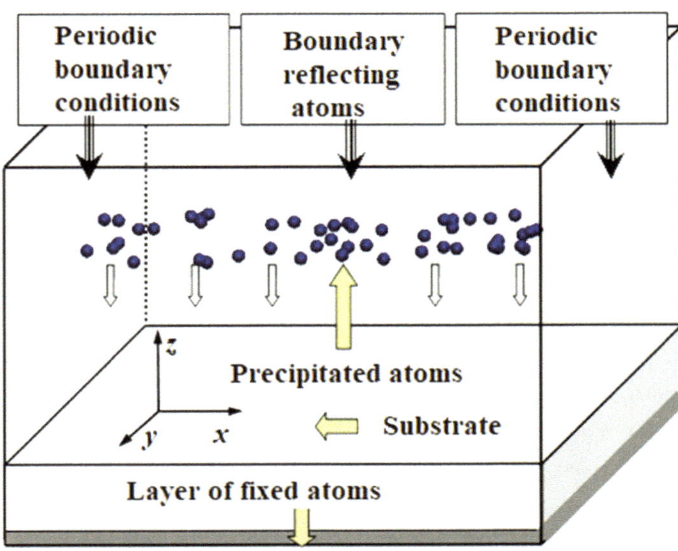

FIGURE 8.2 Calculation scheme for the problem of the formation of superconductor-ferromagnet nanoheterostructures.

As a software tool for conducting theoretical research, the LAMMPS (large-scale atomic/molecular massively parallel simulator) computing complex was used [35]. This software and tool package is freely distributed, contains the ability to perform parallel computing, and supports multilevel mathematical models, including molecular dynamics.

The results were visualized through a separate software tool using VMD (visual molecular dynamics) [36]. Results analysis algorithms were described in TCL and C++ using an additional programming console.

8.3 RESEARCH RESULTS AND ANALYSIS

The formation of multilayer nanofilms was carried out in stages. The first layer was composed of cobalt atoms deposited on a layer of niobium, which plays the role of a substrate. The substrate was a crystalline structure with a size of 13.2 nm in each horizontal direction and a thickness of 3.7 nm.

When plotting the graphs, the upper boundary of the substrate, which is one of the superconductor-ferromagnet contact zones, was taken as a zero height mark. The number of niobium atoms in the substrate was 33,600.

In the first series of computational experiments, the formation of heterostructures occurred on the substrate at a temperature close to normal—300 K. The temperature of the substrate was maintained using a Nose-Hoover thermostat. Note that the temperature of the deposited atoms was not corrected by the thermostat since a directed velocity was applied to them, through which the intensity of the formation of the nanofilm was varied.

In accordance with the technical characteristics from experimental studies [19], in the nano heterostructure, it was required to form the first layer of cobalt with a thickness of 1.5 nm. To achieve this thickness, 18,000 atoms were deposited on the substrate.

The simulation result when spraying the first layer is illustrated in Figure 8.3.

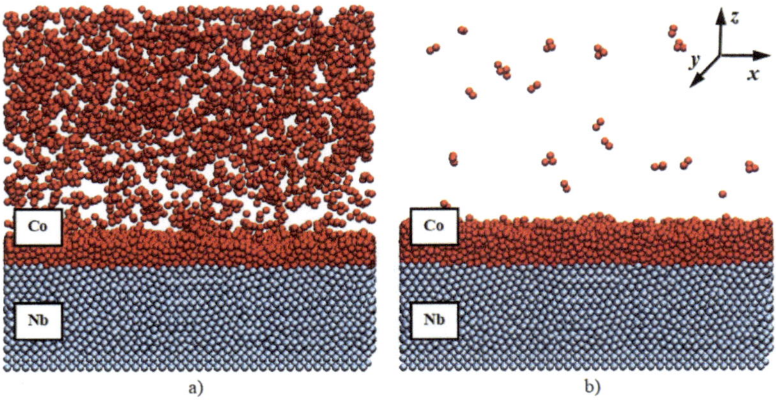

FIGURE 8.3 The pictures of the deposition of the first layer of cobalt on a niobium substrate at a temperature of 300 K, simulation time: (a) 0.1 ns, (b) 0.2 ns.

Here and below, the simulation time in the figures is indicated from the beginning of the deposition of the corresponding layer. The process of formation of the second heterostructure layer is shown in Figure 8.4. The second layer was formed by 70,000 niobium atoms. To achieve the required nanofilm thickness of 8 nm (according to Ref. [19]), a longer simulation time was required.

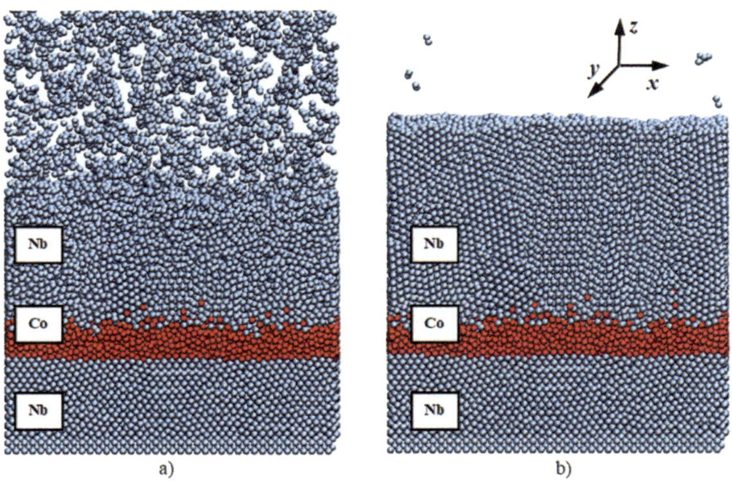

FIGURE 8.4 The pictures of the deposition of a niobium nanofilm on a niobium substrate and a first cobalt layer at a temperature of 300 K, spraying time (a) 0.2 ns, (b) 0.6 ns.

The third heterostructure layer, formed by 30,000 cobalt atoms, is shown in Figure 8.5. After each deposition step, a small number of atoms remained above the sample surface and did not reach the film surface. After completion of the process of layer formation, the required thickness, these atoms were removed from the system.

We note that for all stages of the simulation, the combination of deposited atoms into large conglomerates before contact with the substrate or with the upper layer of the nanofilm was not observed. Figures 8.3–8.5 show the pictures of the processes of nanoheterostructure formation from niobium and cobalt and the morphology of all layers of the system and their interfaces.

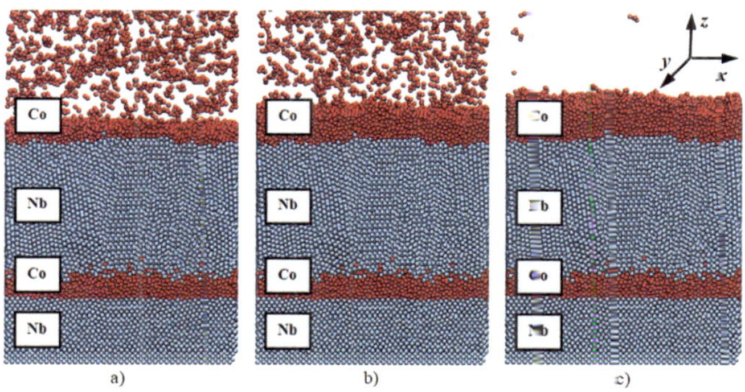

FIGURE 8.5 The pictures of the deposition of the third nanofilm in a heterostructure of niobium and cobalt at a temperature of 300 K, spraying time (a) 0.1 ns, (b) 0.2 ns, and (c) 0.4 ns.

The change in the atomic composition of the layers of the nanosystem is shown in Figures 8.6 and 8.7. These figures are diagrams showing the change in the number of niobium and cobalt atoms in the direction perpendicular to the layers of the nanosystem. As the ordinate on the graphs, the fraction of atoms of a certain type (niobium or chromium) in thin layers of thickness 0.1, parallel to the formed layers of the nanosystem, is indicated. Round markers characterize the content of niobium, square markers-the content of cobalt. The graphs show the change in the fraction of atoms in each layer relative to their total number in this layer. The oblique lines show the graph discontinuity since no changes in the atomic composition in this section of the layer are observed.

The deposition of three layers corresponding to alternating atomic compositions led to the formation of three contact layers in the nanocomposite—superconductor-ferromagnet interfaces. The contact areas in the coordinates H = 0.0 nm (upper plane of the substrate), H = 1.5 nm, and H = 9.5 nm in Figures 8.6 and 8.7 are characterized by a change in structure and the presence of a mixed composition of atoms.

The first contact region (H = 0) between niobium and cobalt has the smoothest contact line and has the smallest mixing of niobium and cobalt atoms since the formation of the first nanofilm occurs on a flat surface of the substrate. The formation of the following layers is carried out on a relief surface formed at the previous stage, the formation of a nanosystem. Therefore, the contact lines between the second and third (H = 1.5 nm), third and fourth (H = 9.5 nm) nanofilms are uneven, and the contact regions containing atoms of different types are more extended. Mixing of atoms of various types in the contact regions is also clearly seen in earlier Figures 8.4 and 8.5.

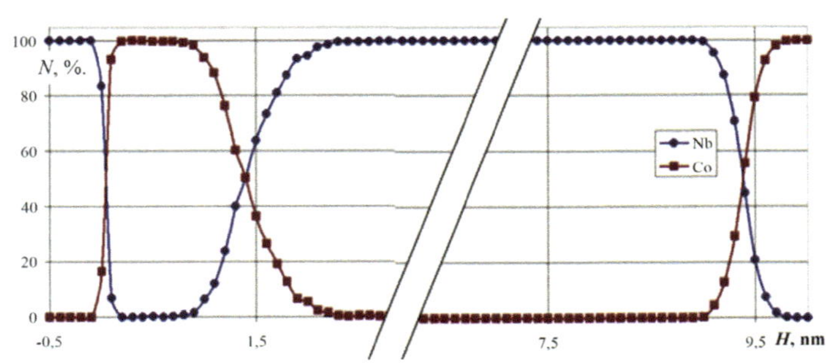

FIGURE 8.6 The change in atomic composition in percent of a nanosystem consisting of four alternating layers of niobium and cobalt formed at a temperature of 300 K.

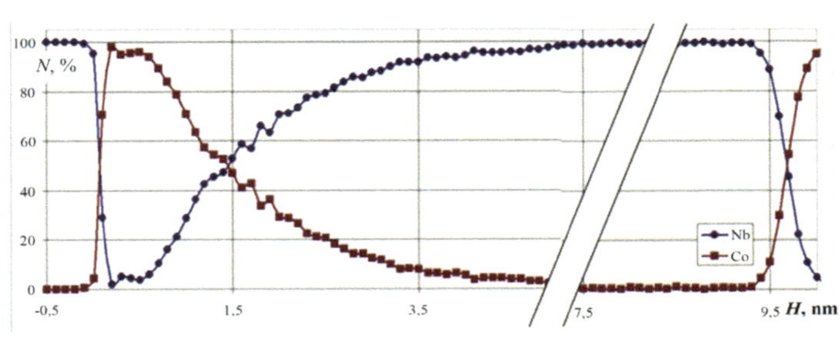

FIGURE 8.7 The change in atomic composition in percent of a nanosystem consisting of four alternating layers of niobium and cobalt formed at a temperature of 800 K.

An analysis of the atomic structure of the layers of the nanosystem shows that niobium and cobalt form different structures. The niobium layer is formed by crystalline atomic regions of different spatial orientations (Figure 8.4, the first and second layers of niobium). The formation of crystallites in time occurs gradually as the niobium layer forms. The cobalt nanofilms are characterized by an atomic structure close to amorphous (Figure 8.4, first and second cobalt layers).

Two more series of computational experiments were carried out in which the formation of similar heterostructures was studied at temperatures of 500 K and 800 K. Quantitative estimates of the structure of materials indicate differences in the processes of nanofilm growth for different temperatures. An increase in temperature to 800 K led to the formation of a structure of less dense samples. At 800 K and 300 K, the difference in the total film thickness of niobium and cobalt was 0.3 nm, which accumulated gradually with each layer. Also, with increasing temperature, the zone of contact areas containing both cobalt and niobium increased, which is clearly seen in the graphs shown in Figure 8.7.

The results obtained indicate a significant dependence of the processes of formation of multilayer nanofilms, the structure of contact areas at the niobium-cobalt boundary, and also the composition, and structure of the temperature at which the formation of a multilayer nanosystem occurs.

8.4 CONCLUSIONS

The paper describes the methodology and theoretical foundations of the study of multilayer heterostructures of the type superconductor-ferromagnet. The analysis algorithms make it possible to study in detail the processes of the formation of nanofilms, identify their structure and composition, identify the imperfection of the structure of samples, and also predict the quality of the interface of contact layers prepared on the basis of the selected materials for deposition.

A study of the structure of nanofilms showed that niobium is formed by crystalline regions of different orientations. The formation of crystallites occurs with a temporary lag from the surface layer, which is associated with the rearrangement of atoms and their desire to occupy an energetically more favorable state after contact with a solid surface. A cobalt nanofilm is characterized by a structure close to amorphous.

The structural features of contact layers of a superconductor-ferromagnet largely depend on the relief of the surface onto which the deposition is carried out. The smallest variation in composition is observed in the first niobium-cobalt contact zone since the formation of the first nanofilm occurs on an even plane of the substrate. The remaining layers begin to precipitate on the already formed relief profile of the solid material and, as a result, have longer contact areas.

An analysis of the influence of the temperature regime during the deposition of a nanosystem indicates the dependence of the processes of formation of multilayer nanofilms, their contact areas, as well as the composition and morphology of heterostructures on the temperature at which the nanocomposite is manufactured. An increased temperature leads to the formation of a structure of samples that are more rarefied and an increase in the zones of mixed contact regions due to the interdiffusion of atoms of the sprayed materials.

Theoretical studies of the fundamental processes of formation and physical properties of multilayer superconducting heterostructures, including a description of the methodology and a comprehensive analysis of the interface layer of the superconductor-ferromagnet materials, allow a detailed study of the spatial profile of the contact region, structural features, and defects in the adjacent region. The identification and description of topological quantum phenomena in multilayer nanosystems will find application in the development of new promising spin devices.

The obtained simulation results can be used in developments as well as optimization of technologies for the formation of spin valves and other functional elements for spintronics.

ACKNOWLEDGMENTS

The works were carried out with financial support from the Russian Science Foundation Grant (RSF) Nr. 20-62-47009 "Physical and engineering basis of computers non-von Neumann architecture based on superconducting spintronics" (Optimization of processes of deposition of multilayer nanosystems by methods of mathematical modeling); the project 0427-2019-0029 Ural Branch of the Russian Academy of Sciences "Study of the laws of formation and calculation of macroparameters of nanostructures and metamaterials based on them using multilevel mathematical modeling" (Modeling the processes of deposition of multilayer nanosystems); the European Union H2020-project "SPINTECH" under grant agreement Nr. 810144 (Experimental studies of deposition multilayer nanosystems).

KEYWORDS

- **mathematical modeling**
- **modified embedded-atom method**
- **molecular dynamics**
- **nanostructure**
- **spin valve**
- **vacuum deposition**

REFERENCES

1. Krupa, M. M., (2007). Spintronics. Problems and prospects of practical application. *Advantage Engineering Technologies, 2*, 1–9.
2. Endoh, T., & Honjo, H., (2018). A recent progress of spintronics devices for integrated circuit applications. *Journal of Low Power Electronics and Applications, 8*(44), 1–17.
3. Chang, C., Kostylev, M., & Ivanov, E., (2013). Metallic spintronic nanofilm as a hydrogen sensor. *Applied Physics Letters, 102*(14), 1–13.
4. Deminov, R. G., Useinov, N. K., & Tagirov, L. R., (2014). Magnetic and superconducting heterostructures in spintronics. *Magnetic Resonance in Solids. Electronic Journal, 16*(2), 14209, 1–9.
5. Bell, C., Burnell, G., Leung, C. W., Tarte, E. J., Kang, D. J., & Blamire, M. G., (2004). Controllable Josephson current through a pseudo-spin-valve structure. *Applied Physics Letters, 84*(7), 1153–1155.
6. Hai, P. N., Ohya, S., Tanaka, M., Barnes, S. E., & Maekawa, S., (2009). Electromotive force and huge magnetoresistance in magnetic tunnel junctions. *Nature, 458*, 489–492.
7. Hai, P. N., & Tanaka, M., (2015). Memristive magnetic tunnel junctions with MnAs nanoparticles. *Applied Physics Letters, 107*, 122404, 1–5.
8. Branford, W. R., Ladak, S., Read, D. E., Zeissler, K., & Cohen, L. F., (2012) Emerging chirality in artificial spin ice. *Science*, *335*(6076), 1597–1600.
9. Dion, T., Arroo, D. M., Yamanoi, K., Kimura, T., Gartside, J. C., Cohen, L. F., Kurebayashi, H., & Branford, W. R., (2019). Tunable magnetization dynamics in artificial spin ice via shape anisotropy modification. *Physical Review B, 100*(5), 054433, 1–11.

10. Uchida, K., Takahashi, S., Harii, K., Ieda, J., Koshibae, W., Ando, K., Maekawa, S., & Saitoh, E., (2008). Observation of the spin see beck effect. *Nature, 455*, 778–781.
11. Jooss, C., Wu, L., Beetz, T., Klie, R. F., Beleggia, M., Schofield, M. A., Schramm, S., et al., (2007). Polaron melting and ordering as key mechanisms for colossal resistance effects in manganites. *Proceedings of the National Academy of Sciences, 104*(34), 13597–13602.
12. Hoffmann, J., Moschkau, P., Mildner, S., Norpoth, J., Jooss, C., Wu, L., & Zhu, Y., (2014). Effects of interaction and disorder on polarons in colossal resistance manganite $Pr_{0.68}Ca_{0.32}$ MnO_3 thin films. *Materials Research Express, 1*(4), 046403, 1–25.
13. Gaifullin, R. R., Kushnir, V. N., Deminov, R. G., Tagirov, L. R., Kupriyanov, M. Y., & Golubov, A. A., (2019). Proximity effect in a superconducting triplet spin-valve S1/F1/S2/F2. *Physics of the Solid State, 61*(9), 1535–1538.
14. Zhaketov, V. D., Nikitenko, J. V., Hajdukov, J. N., Skrjabina, O. V., Chik, A., Borisov, M. M., Muhamedzhanov, J. H., et al., (2019). Magnetic and superconducting properties of inhomogeneous layered structures V/$Fe_{0.7}$ $V_{0.3}$/V/$Fe_{0.7}$ $V_{0.3}$/Nb and Nb/$Ni_{0.65}$($_{0.81}$)$Cu_{0.35}$ (0.19), *156*(2(8)), 310–330.
15. Sidorenko, A. C., (2017). Reentrance phenomenon in superconductor/ferromagnet nanostructures and their application in superconducting spin valves for superconducting electronics. *Low-Temperature Physics*, *43*(7), 766–771.
16. Fominov, Y. V., Golubov, A. A., Karminskaya, T. Y., Kupriyanov, M. Y., Deminov, R. G., & Tagirov, L. R., (2010). Superconducting triplet spin valve. *Journal of Experimental and Theoretical Physics Letters, 91*(6), 308–313.
17. Fominov, Y. V., Golubov, A. A., & Kupriyanov, M. Y., (2003). Triplet proximity effect in FSF trilayers. *Journal of Experimental and Theoretical Physics Letters, 77*(9), 510–515.
18. Zdravkov, V. I., Lenk, D., Morari, R., Ullrich, A., Obermeier, G., Müller, C., Krug, V. N. H. A., et al., (2013). Memory effect and triplet pairing generation in the superconducting exchange biased Co/CoO_x/$Cu_{41}Ni_{59}$/Nb/$Cu_{41}Ni_{59}$ layered heterostructure. *Applied Physics Letters, 103*(6), 62604, 1–9.
19. Zdravkov, V. I., Kehrle, J., Obermeier, G., Lenk, D., Krug, V. N. H. A., Müller, C. K. M. Y., Sidorenko, A. S., et al., (2013). Experimental observation of the triplet spin-valve effect in a superconductor-ferromagnet heterostructure. *Physical Review B, 87*, 144507, 1–6.
20. Vad, K., Csík, A., & Langer, G., (2009). Secondary neutral mass spectrometry-a powerful technique for quantitative elemental and depth profiling analyses of nanostructures. *Spectroscopy Europe, 21*(4), 13–16.
21. Vdovichev, S. N., Nozdrin, J. N., Pestov, E. E., Junin, P. A., & Samohvalov, A. V., (2016). Phase transitions in hybrid SFS structures with thin superconducting layers. *Journal of Experimental and Theoretical Physics Letters, 104*(5), 329–333.
22. Klenov, N., Khaydukov, Y., Bakurskiy, S., Morari, R., Soloviev, I., Boian, V., Keller, T., et al., (2019). Periodic Co/Nb pseudo spin valve for cryogenic memory. *Beilstein Journal of Nanotechnology, 10,* 833–839.
23. Stamopoulos, D., Aristomenopoulou, E., & Lagogiannis, A., (2014). Co/Nb/Co trilayers as efficient cryogenic spin valves and supercurrent switches: The relevance

to the standard giant and tunnel magnetoresistance effects. *Superconductor Science and Technology, 27*(9), 095008, 1–13.

24. Vakhrushev, A. V., (2017). Computational multi-scale modeling of multiphase nanosystems. In *Theory and Applications* (p. 402). Waretown, New Jersey, USA: Apple Academic Press.
25. Vakhrushev, A. V., & Fedotov, A. Y., (2007). Modeling of composite nanoparticle formation from a gas phase. *International Scientific Journal Alternative Energy and Ecology, 10*(54), 22–26.
26. Vakhrushev, A. V., Fedotov, A. Y., Vakhrushev, A. A., Golubchikov, V. B., & Givotkov, A. V., (2011). Multilevel simulation of the processes of nanoaerosol formation. Part 2. Numerical investigation of the processes of nanoaerosol formation for suppression of fires. *International Journal of Nanomechanics Science and Technology, 2*(3), 205–216.
27. Suyetin, M. V., & Vakhrushev, A. V., (2009). Nanocapsule for safe and effective methane storage. *Nanoscale Research Letters, 4*(11), 1267–1270.
28. Volkova, E. I., Suyetin, M. V., & Vakhrushev, A. V., (2010). Temperature-sensitive nanocapsule of complex structural form for methane storage. *Nanoscale Research Letters, 5*(1), 205–210.
29. Suyetin, M. V., & Vakhrushev, A. V., (2011). Guided carbon nanocapsules for hydrogen storage. *Journal of Physical Chemistry C, 115*(13), 5485–5491.
30. Vorobiev, V. L., Bykov, P. V., Bayankin, V. Y., Shushkov, A. A., Vakhrushev, A. V., & Orlova, N. A., (2012). Mechanical properties of carbon steel changes with the beam mean current density under pulsed irradiation with Ar ions. *Physics and Chemistry of Materials Processing, 6,* 5–9.
31. Vakhrushev, A. V., Severyukhina, O. Y., Severyukhin, A. V., Vakhrushev, A. A., & Galkin, N. G., (2012). Simulation of the processes of formation of quantum dots on the basis of silicides of transition metals. *International Journal of Nanomechanics Science and Technology, 3*(1), 51–75.
32. Valeev, R. G., Vakhrushev, A. V., Fedotov, A. Y., & Petukhov, D. I., (2019). *Functional Semiconductor Nanostructures in Porous Anodic Alumina Matrices: Modeling, Synthesis, Properties* (p. 285). Waretown, USA: Apple Academic Press.
33. Baskes, M. I., (1992). Modified embedded-atom potentials for cubic materials and impurities. *Physical Review B, 46*(5), 2727–2742.
34. Lee, B. J., Baskes, M. I., Kim, H., & Cho, Y. K., (2001). Second nearest-neighbor modified embedded atom method potentials for bcc transition metals. *Physical Review B, 64*(18), 184102, 1–11.
35. Plimpton, S., (1995). Fast parallel algorithms for short-range molecular dynamics. *Journal of Computational Physics, 117*(1), 1–19.
36. Humphrey, W., Dalke, A., & Schulten, K., (1996). VMD: Visual molecular dynamics. *Journal of Molecular Graphics, 14*(1), 33–38.

CHAPTER 9

Modeling of Deformation and Fracture Process of Layered Nanocomposites

A. YU. FEDOTOV,[1,2] A. T. LEKONTSEV,[2] and A. V. VAKHRUSHEV[1,2]

[1]*Department of Mechanics of Nanostructures, Institute of Mechanics, Udmurt Federal Research Center, Ural Division, Russian Academy of Sciences, Izhevsk, Russia*

[2]*Department of Nanotechnology and Microsystems, Kalashnikov Izhevsk State Technical University, Izhevsk, Russia, E-mail: Vakhrushev-a@yandex.ru*

ABSTRACT

The chapter formulates the problem of deformation and fracture of nanocomposites by the molecular dynamics method. To describe the interatomic potential used embedded atom. Molecular dynamic modeling of uniaxial tension of a layered Al/Cu nanocomposite has been performed. Deformation nanocomposite is carried by elastic and plastic deformation of the material to fracture. The basic parameters of the deformation of materials are investigated-deformation, stress, temperature, and atomic structure of the material. Modeling showed that when the stresses in the sample reached the elastic limit, nucleation of defects in the crystal lattice of the material and their propagation through the crystal in the form of shifts and rotations of atoms in the crystal planes were observed. The areas of nucleation of plastic strains and the formation of defects are determined. The maximum destruction of the material occurred at the interface of the components of the nanocomposite.

9.1 INTRODUCTION

The interest in studying the behavior of composite materials under loading has a set of properties and features that differs from traditional structural materials. These properties can be high strength or low density of the material, as well as the ability to change the physical, chemical, or mechanical characteristics during production. In addition, the question of the distribution of stresses and strains in nanomaterials is of interest.

Currently, in the development of models for describing the mechanisms of deformation occurring in a material, the properties and parameters of materials and computer modeling are widely used. Modeling shows that the interface between materials with different mechanical characteristics is sources of stress and can become additional sources of defects [1, 2]. In Ref. [3], the process of uniaxial tension with a constant velocity along the interface of a sample consisting of aluminum and nickel crystallites was considered by the molecular dynamics method. It was found that when the system reaches the elastic limit at the interfaces, crystal lattice defects arise, which subsequently propagate through the material. In Ref. [4], the influence of the microstructure of a material on its plastic deformation and its destruction during high-speed deformation, which occurs, for example, in shock wave phenomena [5], is studied. In Ref. [6], the mechanisms of nucleation and growth of dislocation loops in a defect-free crystal under uniaxial compression strains and pure shear are studied. This work also evaluated critical stresses for nucleation of dislocation loops. The study of the mechanisms of deformation and fracture was carried out on aluminum using molecular dynamics modeling. For the description of the processes used capacity of the embedded atom (EAM).

In general, under shear stresses, elastic and plastic deformation, defects in the crystal lattice, and damage occur in the crystal structure of the material. These phenomena that occur on an atomic scale are complex for experimental studies. The most difficult to investigate the described processes directly in the process of loading [7–10].

In this work, the method of complex loading of a layered nanocomposite is studied by the molecular dynamics method. At the first stage, the material was cooled, at the second; the material was stretched in the uniaxial direction at a constant speed along with the interface of a sample consisting of plane cooled crystallites (defect-free single crystals) of aluminum and copper. The choice of cooled nanocomposites for studying the deformation processes is due to the need for additional studies of nanomaterials used at low temperatures, for example, topological insulators [11, 12]. This work is a continuation and development of research on modeling of deformation processes of nanocomposites with fillers of various shapes [14–18].

9.2 MATHEMATICAL MODEL AND STATEMENT OF THE PROBLEM

For the simulation, we used the free software package for classical molecular dynamics LAMMPS (large-scale atomic/molecular massively parallel simulator). To visualize the results obtained, the Ovito software package was used. The well-established embedded atom method (EAM) was chosen as the potential of interatomic interaction. Let us briefly describe this method.

This potential explicitly takes into account the influence of the electronic subsystem on the interaction of atoms with each other in metals [19, 20]. The electron density ρ_i at some point r_i where the atom i is located is written in the form of separate electron densities ρ_{ij} created by other atoms j:

$$\rho_i = \sum_{j \neq i}^{N_c} \rho_{ij}(r_{ij}) \tag{1}$$

where, N_c is the number of atoms enclosed in the cutoff with a radius r_{cut}; r_{ij} is the distance between atoms, which is equal to $\sqrt{\sum_\alpha \left(x_{i\alpha} - x_{j\alpha}\right)^2}$.

After averaging, it is assumed that the electron density ρ_{ij} depends only on the distance between the atoms r_{ij}. Since the electron density decreases with increasing distance from the nucleus, the following formula is used for approximation:

$$\rho_{ij}\left(r_{ij}\right) = \rho_{ij}^{0} \exp\left(-\beta\left(\frac{r_{ij}}{r_e} - 1\right)\right) \tag{2}$$

where r_e is the equilibrium distance between nearest neighboring atoms.

The energy with which ions interact with the electron field is represented as a function $F_i\rho_i$ that depends on the state of electron density.

Ions are attracted to each other due to their interaction with the electron cloud, which is located between them. At first, the EAM potential was used to describe and model metals with a face-centered cubic lattice and a body-centered cubic lattice. Subsequently, this potential was improved for metals with a hexagonal close-packed crystal lattice.

The disadvantage of the EAM potential is that it does not take into account the direction of chemical bonds that arise when nuclei interact with π-electrons. Nevertheless, in most cases, the potential gives a satisfactory result when reproducing various properties of a wide range of chemical elements. The potential energy acting on a specific atom, calculated by the EAM potential method, is written in the following form:

$$V_i = F_i(\rho_i) + \frac{1}{2}\sum_{j\neq i}^{N_c} \varphi(r_{ij}) \tag{3}$$

The ion repulsion due to the pair Coulomb interaction $\varphi(r_{ij})$ is determined by the second term in formula (3).

A detailed description of the immersed atom EAM method as applied to the problems of the deformation and destruction of composites can be found in Ref. [19].

To carry out the calculations, the Nose-Hoover thermostat was used [20]. In the thermostat, a thermal reservoir and friction losses are added to the system. The force of friction between atoms is proportional to the product of the speed of the atom and the coefficient of friction ξ. The value of the time derivative ξ is determined by the difference between the current kinetic energy and the energy value corresponding to the initial temperature:

$$\frac{d^2 r_i}{dt^2} = \frac{f_i}{m_i} - \xi\frac{dr_i}{dt}, i = \overline{1,N}, \frac{d\xi}{dt} = \frac{1}{Q}(T - T_{ext}), Q = \frac{\tau_T^2 T_{ext}}{4\pi^2} \tag{4}$$

where, Q is the mass ratio; T_{ext} is the thermostat setting temperature; T is the current system temperature; and τ_T is the atomic oscillation period.

Let us consider in more detail the calculation of deformations and stresses in a nanocomposite. To calculate the stress and strain tensors in nanosystems, several approaches are currently used. Methods for calculating these tensors have a different physical nature and rely on alternative mathematical tools. Consider two basic methods in more detail.

1. **The First Approach:** It is based on the relationships of continuum mechanics [21] and provides for the construction of the calculation process in the "top-down" direction. At the preliminary stage of calculating the elastic characteristics, an elementary volume is allocated, in relation to which these characteristics will be related. There are certain difficulties with the calculation of the elementary volume in nanomaterials since there is no unambiguous formulation of the volume of an individual atom. In addition, between the atoms in the system, there are voids, inhomogeneities, and dislocations, which also need to be included in the volume of the sample under study. Incorrect consideration of intercrystalline faces can lead to inaccurate and erroneous calculations of stress and strain tensors. On the one hand, an increase in the size of the sample makes it possible to reduce the magnitude of the error; on the other hand, based on the definitions of continuum mechanics, the volume should remain infinitely small.

Having determined the value of the elementary volume, calculate the area of the planes bounding it. For definiteness, we denote the area of $S_{\alpha\beta}$ borders. Hereinafter, the indices α and β indicate the direction and position of the boundary, for a three-dimensional Cartesian coordinate system $\alpha, \beta = \{1,2,3\}$. Some facets of the volume are applied forces $F_{\alpha\beta}$ that can act tangentially and normal, therefore, when designating forces, there are also indices α and β. Under the action of forces, the nano-volume begins to deform, its geometry changes.

The components of the stress tensor are calculated through the corresponding forces and areas of the boundaries of the nanomaterial

$$\sigma_{\alpha\beta} = \frac{F_{\alpha\beta}}{S_{\alpha\beta}}. \quad (5)$$

In the absence of an intrinsic angular momentum of a continuous medium, as well as bulk and surface stress pairs, the stress tensor is symmetric, that is $\sigma_{\alpha\beta} = \sigma_{\beta\alpha}$, and has six independent components.

The vector of displacements of the elementary volume is determined through the averaged displacements of all atoms using the values of the initial and final coordinates in accordance with the ratio

$$\mathbf{u} = \frac{1}{N}\sum_{k=1}^{N}\mathbf{u}_k = \frac{1}{N}\sum_{k=1}^{N}\mathbf{r}_k - \mathbf{r}'_k = \frac{1}{N}\sum_{k=1}^{N}\Delta\mathbf{r}_k \text{ ,} \tag{6}$$

where $\mathbf{u}$ is the vector of displacements of the elementary volume; $\mathbf{u}_k$ is the displacement vector for the k-th atom; N is the number of atoms in the nanosystem; $\mathbf{r}_k$ and $\mathbf{r'}_k$ are the radius vectors of each atom before and after the movement, respectively.

The strain tensor, which is responsible for changing the shape and rebuilding the size of a nanomaterial, is associated with displacements by the expression:

$$\varepsilon_{\alpha\beta} = \frac{1}{2}\left(\frac{\partial u_\alpha}{\partial r_\beta} + \frac{\partial u_\beta}{\partial r_\alpha} + \sum_l \frac{\partial u_1}{\partial u_\alpha}\frac{\partial u_1}{\partial r_\beta}\right), \tag{7}$$

where $\varepsilon_{\alpha\beta}$ are the components of the strain tensor; u_α, u_β, r_α, r_β are the vector components of displacement and vector radius. Based on the definition, the strain tensor is also asymmetric quantity.

For small displacements, the summation over the index l neglects and uses the strain tensor in a simpler form

$$\varepsilon_{\alpha\beta} = \frac{1}{2}\left(\frac{\partial u_\alpha}{\partial r_\beta} + \frac{\partial u_\beta}{\partial r_\alpha}\right) \tag{8}$$

The calculation of the partial derivatives in Eqns. (7) and (8) involves a number of difficulties since it is not possible to obtain these expressions explicitly. Derivatives are determined numerically, which affects the accuracy and adequacy of the resulting strain tensor.

Using the tenors of stress and strain, Hooke's law can be written and various elastic properties of a nanomaterial such as Young's modulus, Poisson's ratio, shear modulus, and volume expansion coefficient can be calculated. In generalized form, Hooke's law is written as

$$\sigma_{\alpha\beta} = \sum_{k,l} C_{\alpha\beta kl} \cdot \varepsilon_{kl} \text{ ,} \tag{9}$$

where $C_{\alpha\beta kl}$ is the fourth rank tensor. For an isotropic material, the tensor $C_{\alpha\beta kl}$ contains only two independent coefficients.

2. **The Second Approach:** It is an alternative, provides for the construction of the calculation process in the "bottom-up" direction, and relies on the kinetic theory and the virial theorem [22, 23]. The virial theorem connects the average kinetic energy of a nanosystem with the average potential energy and allows us to take into account not only the component of the molecular-kinetic theory of an ideal gas but also the influence of the properties of real atoms:

$$P = \frac{Nk_BT}{W} + \frac{1}{3W}\sum_{k=1}^{N'} \mathbf{r}_k \cdot \mathbf{F}_k, \tag{10}$$

where W is the volume of the settlement area; $\mathbf{F}_k$ si the resultant force acting on the k-th atom; $\mathbf{r}_k$ is the radius vector of the k-th atom. The value N' is responsible for the number of atoms, taking into account the symmetric images used in the simulation with periodic boundary conditions. The second term in Eq. (10) is virial.

The components of the nanosystem pressure tensor are calculated in a similar way. Given the relationship between the kinetic and potential energies and coordinate wise decomposition of the vectors, the expression for the pressure tensor is used.

$$P_{\alpha\beta} = \frac{1}{W}\sum_{k=1}^{N} m_k V_{k,\alpha} V_{k,\alpha} + \frac{1}{W}\sum_{k=1}^{N'} r_{k,\alpha} \cdot F_{k,\beta}, \tag{11}$$

where m_k is the mass of the k-th atom; $V_{k,\alpha}$, $V_{k,\beta}$ are the components of the velocity of the k-th atom; $r_{k,\alpha}$, $F_{k,\beta}$ are the elements of the radius vector and force for the k-th atom, respectively.

Due to the dependence of the force on the potential gradient, the stress-strain state will ultimately also be determined by the potential field. For an individual atom under the number k, the analog of the stress tensor is written as the following expression:

$$\begin{aligned} f_{k,\alpha\beta} = -\Bigg[& m_k V_{k,\beta} + \frac{1}{2}\sum_{n=1}^{N_p}\left(r_{1,\alpha}F_{1,\beta} + r_{2,\alpha}F_{2,\beta}\right) + \frac{1}{2}\sum_{n=1}^{N_b}\left(r_{1,\alpha}F_{1,\beta} + r_{2,\alpha}F_{2,\beta}\right) \\ & + \frac{1}{3}\sum_{n=1}^{N_a}\left(r_{1,\alpha}F_{1,\beta} + r_{2,\alpha}F_{2,\beta} + r_{3,\alpha}F_{3,\beta}\right) + \frac{1}{4}\sum_{n=1}^{N_d}\left(r_{1,\alpha}F_{1,\beta} + r_{2,\alpha}F_{2,\beta} + r_{3,\alpha}F_{3\beta} + r_{4,\alpha}F_{4\beta}\right) + \\ & + \frac{1}{4}\sum_{n=1}^{N_i}\left(r_{1,\alpha}F_{1,\beta} + r_{2,\alpha}F_{2,\beta} + r_{3,\alpha}F_{3,\beta} + r_{4,\alpha}F_{4,\beta}\right) + Kspace\left(r_{k,\alpha}, F_{k,\beta}\right) + \sum_{n=1}^{N_f} r_{k,\alpha}F_{k,\beta} \Bigg] \end{aligned} \tag{12}$$

Formula (12) takes into account the different types of potentials and interactions and the kinetic energy of the atom. The first term is the contribution of kinetic energy to the force tensor; therefore, the product of the components of velocity is present in it. The second term of the sum is responsible for the pair interaction, where N_p is the number of the nearest neighbors of the atom *k* participating in the pair interaction, and $\mathbf{F}_1$, $\mathbf{F}_2$ are the forces that arise. The third term corresponds to the nodal potential, therefore, under the sum sign, there are the products of three radius vectors and forces, N_a is the number of angles. Similar to the previous elements, N_d and N_i are the number of dihedrals and pseudo-double-angled angles, the interaction is calculated by four atoms. The term *Kspace* describes the long-range Coulomb contribution of the potential, and the last term relates to fixed and bounded particles, N_f is the number of fixed atoms. In most problems, not all types of interactions are present, therefore, Eq. (13) is significantly simplified.

For an infinitely small volume, the stress tensor is calculated by summing over all atoms:

$$\sigma_{\alpha\beta} = \frac{1}{W_e} \sum_{k=1}^{N} f_{k,\alpha\beta} \,, \tag{13}$$

where W_e is the value of the elementary volume.

Displacements in the second approach are also determined on the basis of formula (9). On the basis of the distortion of the geometry of the elementary volume, for example, during the transition from a parallelepiped to a prismatic form, the strain tensor and the other mechanical parameters of the nanomaterial are found. In this paper, the second approach was used to describe the stress-strain state.

9.3 RESULTS OF SIMULATIONS AND ANALYSIS

We will analyze the simulation results on the problem of the uniaxial tension of a two-layer composite of aluminum and copper.

Modeling was carried out in two stages. At the first stage, a sample consisting of two crystallites of aluminum and copper in the form of parallelepipeds connected along one of the collective boundaries (Figure 9.1) was placed in the computational domain and was cooled to 0 K at constant pressure. At the second stage, the main parameters of the layered crystal were calculated: deformations, internal energy, its kinetic and potential components, temperature distribution, and stress in the crystal during the stretching of the sample. To support pressure and temperature, the algorithms of thermostats and barostats were used in the first stage of modeling. Since the pressure includes the kinetic component due to atomic velocities, both of these algorithms require a temperature calculation. Typically, the target temperature and pressure are set when setting the task, and the thermostat and barostat try to balance the system with the required temperature and pressure.

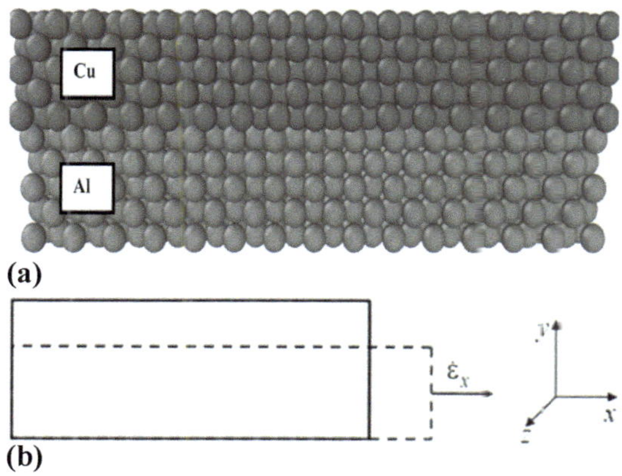

FIGURE 9.1 Image of the sample, consisting of crystallites of aluminum and copper (a) and a tensile diagram of the sample (b).

The dimensions of the composite crystallite along the axes x, y, and z were 60, 24, and 24 angstroms, respectively. The total number of atoms in the nanosystem is 2200. Periodic boundary conditions were used on all surfaces of the crystal. At the initial stage, the computational region was cooled to a temperature close to 0°K. As a result of the cooling, the atoms in the nanosystem stabilized and occupied positions corresponding to the minimum energy value. After balancing the system, the crystal structure of the materials at the interfaces is distorted, and atoms exchange in the intermediate layer between the copper and aluminum atoms. This is due to the mismatch of the lattice constants for aluminum and copper. As a result, the ideal crystal lattices of these materials are distorted, and compression and dislocation stresses appear. Figure 9.2 shows the atomic structure of a cooled sample. In this figure, the brightest atoms have a maximum displacement relative to the equilibrium crystal lattice.

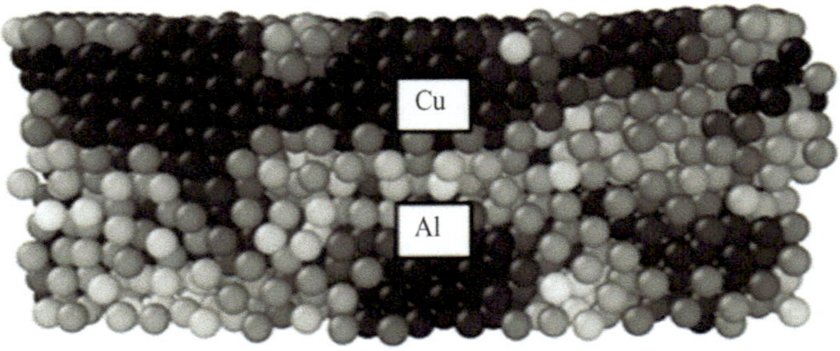

FIGURE 9.2 Atomic structure of the sample at minimum temperature.

In the process of uniaxial tension of the computational cell, the sample first deforms elastically (Figure 9.3, structures 1–2). After the material reaches the elastic limit, a transition to plastic deformation occurs, accompanied by structural changes in the atomic system (Figure 9.3, structures 3–4). In Figure 9.4, plastic deformation corresponds to a plot of the graph along the abscissa after a strain value of 0.12. During elastic deformation in the region, the temperature of the sample is close to 0 K, and the potential energy of the atomic system increases since the elastic strain energy is accumulated in the system.

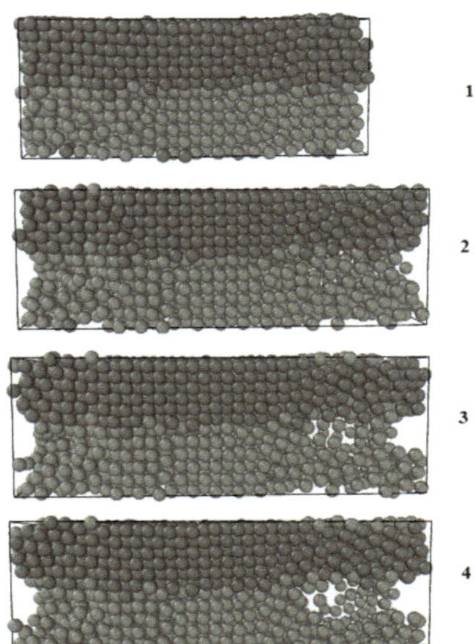

FIGURE 9.3 Change in the atomic structure of the sample during tensile deformation.

During deformation $\varepsilon > \varepsilon_c$ in the sample, we observe the appearance of atomic shifts in the structure of the material, and plastic deformation is formed. In this case, the temperature of the sample begins to increase sharply due to the release of thermal energy, as shown in Figure 9.5.

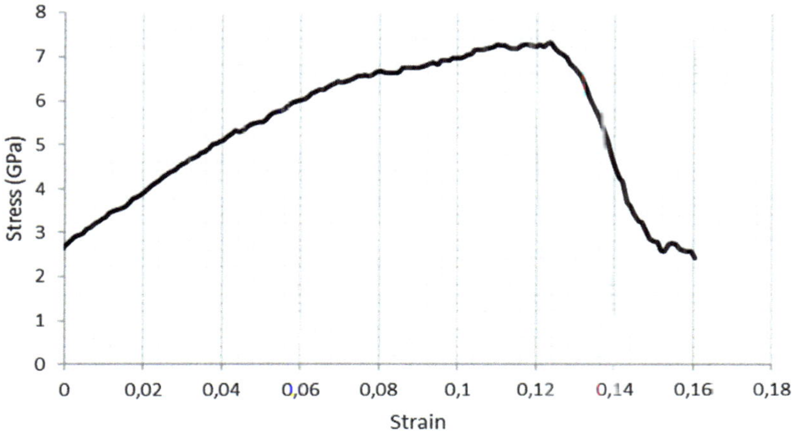

FIGURE 9.4 The stress-strain curve when the sample is stretched.

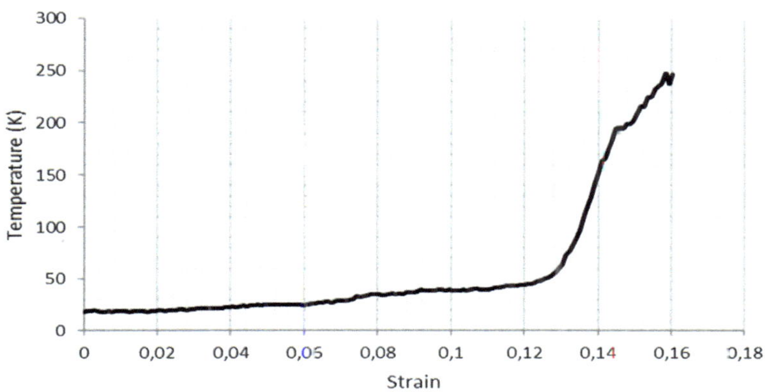

FIGURE 9.5 The dependence of the temperature of the nanocomposite on the degree of deformation of the sample under tension.

In the process of deformation of the crystal lattice, damage propagates in such a way that significant rotations of the atomic planes take place in the volume of aluminum, and Luders-Chernov waves, which propagate through the crystal, are visible in the volume of copper. These waves propagate by shifts in the {111} planes, which are most unstable with respect to the shift in FCC metals in the direction of the highest tangential stress.

An example of the distribution of these waves is shown in Figure 9.6, which shows the atomic structure of a collapsing sample. In this figure, the lightest atoms, as in Figure 9.3, have maximum displacement relative to the equilibrium crystal lattice.

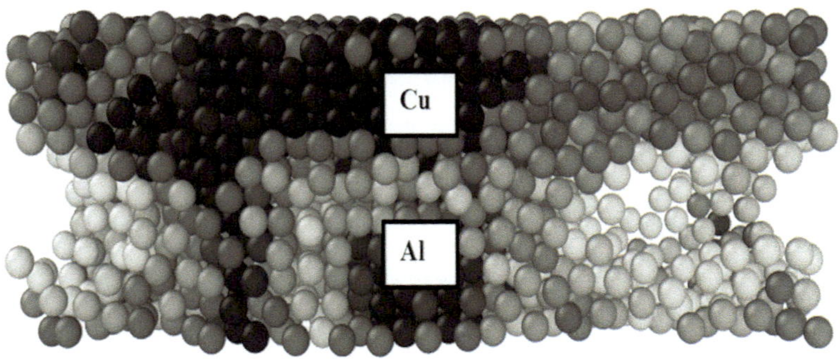

FIGURE 9.6 The atomic structure of the sample at the time of destruction.

Thus, the most convenient and effective way to determine the areas of nucleation of plastic strains is to plot the temperature distribution over the sample during loading. Modeling has demonstrated that in the regions of the onset of defects in the crystal lattice, the temperature can significantly exceed the average temperature of the studied sample. The temperature distributions in the nanocomposite also made it possible to determine the slip planes during plastic shifts. Note that in experimental studies, an increase in temperature in the region of plastic deformation and fracture of the sample is also observed.

9.4 CONCLUSIONS

Molecular dynamics modeling allows us to consider what happens to the material at the nanoscale under loading, to study the propagation and nature of plastic deformation, the formation of defects and damage in the crystal structure of the sample. The used multi-particle interatomic potential of EAM has demonstrated a good reproduction of the properties of the nanocomposite in the study of deformation processes.

An algorithm and simulation script for the nanocomposite deformation problem has been prepared in the LAMMPS software package. The algorithm allows you to dynamically control the studied characteristics of the sample, including strain, thermodynamic, energy and size parameters, stresses, as well as interactively monitor the processes of nucleation of dislocations and destruction of the sample.

The behavior of a cooled Al/Cu nanocomposite under tension with a constant strain rate was studied. During loading of the material, defects in the crystal lattice are formed at the interface, which propagates throughout the volume of the sample. During the deformation of the Al/Cu composition, the regions of nucleation of plastic deformation are determined. The maximum destruction of the material occurs at the interface of materials.

The results of the research can be used to study the deformation processes of nanocomposite materials with advanced functionalities.

ACKNOWLEDGMENTS

The works were carried out with financial support from the Research Program of the Ural Branch of the Russian Academy of Sciences (project 18-10-1-29, project 0427-2019-0029), Kalashnikov Izhevsk State Technical University (project 28.04.01/18BAB).

KEYWORDS

- **defects**
- **deformation**
- **dislocation**
- **mathematical modeling**
- **modified embedded-atom method**
- **molecular dynamics**
- **nanocomposites**

REFERENCES

1. Panin, S. V., Koval, A. V., Trusova, G. V., Pochivalov, Y. I., & Sizova, O. V., (2000). The influence of the geometry and structure of the interface on the nature of the development of plastic deformation at the mesoscale level of borated structural steel samples. *Physical Mesomechanics. T., 3*(2), 99–115.
2. Golovnev, I. F., Golovneva, E. I., Konev, A. A., & Fomin, V. M., (1998). Physical mesomechanics and molecular dynamics modeling. *Physical Mesomechanics, 2*, 21–33.
3. Zhilyaev, P. A., Kuksin, A. Y., Norman, G. E., Starikov, S. V., Stegailov, V. V., & Yanilkin, A. V., (2010). Influence of the microstructure of a material on dynamic plasticity and strength: Molecular-dynamics modeling. *Physics and Chemical Kinetics in Gas Mechanics, 9*(1), 104–109.
4. Bolesta, A. V., Golovnev, I. F., & Fomin, V. M., (2002). Molecular-dynamic modeling of quasistatic stretching of the AL/Ni composition along with the interface. *Physical Mesomechanics,* (4), 15–21.
5. Kanel, G. I., Fortov, V. E., & Razorenov, S. V., (2007). Shock waves in condensed matter physics. *Successes of Physical Sciences, 177*(8), 809–830.
6. Goryacheva, I. G., (2001). *The Mechanics of Frictional Interaction* (p. 478). M: Nauka.
7. Panin, V. E., Yelsukova, T. F., & Grinyaev, Y. V., (2003). The mechanism of the effect of grain size on the resistance to deformation of polycrystals in the concept of structural levels of deformation of solids. Part I. The need to consider mesoscopic structural levels of deformation when analyzing the hall-petch equation. *Physical Mesomechanics, 6*(3), 63–74.
8. Persson, B. N. J., (2002). Elastic contact between randomly rough surfaces: Comparison of theory with numerical results. In: Persson, B. N. J. Bucher, F., & Chiaia, B., (eds.), *Phys. Rev.* (Vol. B65, No. 18, pp. 184106, 1–7).
9. Iordanoff, I., (1999). First steps for a rheological model for the solid third body. *I. Iordanoff, Y. Berthier. Tribology Series, 36*, 551–559.
10. Krivtsov, A. M., Volkovets, I. B., Tkachev, P. V., & Tsaplin, V. A., (2002). Application of the particle dynamics method for describing high-speed fracture of solids. *Transactions of the All-Russian Conference "Mathematics, Mechanics and Computer Science 2002"* (pp. 361–377).
11. Vedeneev, S. I., (2017). Quantum oscillations in three-dimensional topological insulators. *Successes of Physical Sciences, 187*, 411–429.
12. Beg, F., Puzhol, P., & Ramazashvili, R., (2018). Identification of two-dimensional antiferromagnetic topological insulators of class Z2. *ZhETF, 153*(1), 108–126.
13. Vakhrushev, A. V., Fedotov, A. Y., & Vakhrushev, A. A., (2011). Modeling of processes of composite nanoparticle formation by the molecular dynamics technique. Part 1. Structure of composite nanoparticles. *Nanomechanics Science and Technology, 2*(1), 9–38.
14. Vakhrushev, A. V., Fedotov, A. Y., & Vakhrushev, A. A., (2011). Modeling of processes of composite nanoparticle formation by the molecular dynamics technique.

Part 2. Probabilistic laws of nanoparticle characteristics. In: Vakhrushev, A. V., (Ed.), *Nanomechanics Science and Technology,* (Vol. 2, No. 1, pp 39–54).

15. Vakhrushev, A. V., Fedotov, A. Y., & Shushkov, A. A., (2015). Calculation of the elastic parameters of composite materials based on nanoparticles using multilevel models. In: Vladimir, I. K., Gennady, E. Z., & Haghi, A. K., (eds.), *Nanostructures, Nanomaterials, and Nanotechnologies to Nanoindustry* (pp. 51–70). New Jersey: Apple Academic Press, Chapter 4.
16. Vakhrushev, A. V., & Fedotov, A. Y., (2019). Simulation of deformation and fracture processes in nanocomposites. *Fracture and Structural Integrity, 49*, 370–382.
17. Fedotov, A. Y., & Vakhrushev, A. V., (2020). Study of properties of nanostructures and metal nanocomposites on their basis. Nanomechanics and micromechanics. In: Satya, B. S., Alexander, V. V., & Haghi, A. K., (eds.), *Generalized Models and Nonclassical Engineering Approaches.* (pp. 17–42). Waretown, USA: Apple Academic Press.
18. Daw, M. S., (1984). Embedded-atom method: Derivation and application to impurities, surfaces, and other defects in metals. In: Daw, M. S., & Baskes, M. I., (eds.), *Physical Review B* (Vol. 29, No. 12, pp. 6443–6453).
19. Foiles, S. M., (1986). Embedded-atom method function for the FCC metals Cu, Ag, Au, Ni, Pd, Pt, and their alloys. In: Foiles, S. M., Baskes, M. I., & Daw, M. S., (eds.), *Phys. Rev. B* (Vol. 33, No. 12, pp. 7983–7991).
20. Hoover, W., (1985). Canonical dynamics: Equilibrium phase-space distributions. In: Hoover, W., (ed.), *Physical Review A* (Vol. 31, No. 3, pp. 1695–1697).
21. Golovneva, E. I., Golovnev, I. F., & Fomin, V. M., (2005). Peculiarities of application of continuum mechanics methods to the description of nanostructures. *Phys. Mesomech., 8*(5/6), 41–48.
22. Hummer, G., Gronbech-Jensen, N., & Neumann, M., (1998). Pressure calculation in polar and charged systems using Ewald Summation: Results for the extended simple point charge model of water. *Journal of Chemical Physics, 109*(7), 2791, 1–19.
23. Thompson, A. P., Plimpton, S. J., & Mattson, W., (2009). General formulation of pressure and stress tensor for arbitrary many-body interaction potentials under periodic boundary conditions. *J. Chem. Phys., 131*(15), 154107, 1–6.

CHAPTER 10

Elastoplastic Modeling of an Orthotropic Boron-Aluminum Fiber-Reinforced Composite Thick-Walled Rotating Cylinder Subjected to a Temperature Gradient

A. G. TEMESGEN,[1] S. B. SINGH,[1] and T. PANKAJ[2]

[1]*Department of Mathematics, Punjabi University, Patiala, India*

[2]*ICFAI University, Himachal Pradesh, India*

ABSTRACT

The objective of this chapter is to derive the problem of elastoplastic modeling of an orthotropic boron-aluminum fiber-reinforced composite thick-walled rotating cylinder subjected to a temperature gradient by using Seth's transition and generalized strain measure theory. The combined effects of temperature and angular speed have been presented numerically and graphically. Seth's transition theory does not require the assumptions: the yield criterion, the incompressibility conditions, the deformation is small, etc., and thus solves a more general problem. This theory utilizes the concept of generalized strain measure and asymptotic solution at the turning points of the differential equations defining the deformed field. It is seen that cylinders having smaller radii ratios require higher angular speed for yielding as compared to cylinders having higher radii ratios. With the inclusion of thermal effects, the angular speed increased for initial yielding to a smaller radii ratio but for the fully plastic state, the angular speed is the same. It is observed that the maximum circumferential stress occurs at the internal surface for both transitional and fully plastic state at any temperature and angular speed.

10.1 INTRODUCTION

The cylinder is a structural component in various engineering applications such as aerospace industries, pressure vessels, nuclear reactors, military applications, piping, etc. Davis and Conelly [3] studied rotating cylinders and tubes of a strain-hardening material. Thick-walled circular cylinders are used commonly either as pressure vessels intended for the storage of gases or as media transportation of high pressurized fluids.

Many authors [3, 6, 7, 13] discussed problems by classical treatment in an isotropic thick-walled cylinder under internal pressure by using some simplifying assumption like incompressibility of material and yield criterion. The incompressibility of the material is one of the most important assumptions, which simplifies the problem. In fact, in most of the cases, without this assumption, it is impossible to get a closed-form solution.

Gupta and Bhardwaj [5] solved the problem of elastic-plastic and creep transition in an orthotropic rotating cylinder by using Seth's transition and generalized strain measure theory. Seth's transition theory does not require the assumptions: the yield criterion, the incompressibility conditions, the deformation is small, etc., and thus solves a more general problem, from which cases pertaining to the above assumptions can be worked out. Seth's transition theory utilizes the concept of generalized strain measure and asymptotic solution at the turning points or critical points of the differential equations defining the deformed field and has been successfully applied to a larger number of the problems [6, 7, 9].

Seth [10] has defined the generalized principal strain measure as

$$e_{ii} = \int_0^{e_{ii}^A} (1-2e_{ii}^A)^{\frac{n}{2}-1} de_{ii}^A = \frac{1}{n}(1-(1-2e_{ii}^A)^{\frac{n}{2}}) \qquad (1)$$

where, n is the coefficient of strain measure, e_{ii}^A is the principal Almansi finite strain component and $i = 1, 2, 3$.

In this chapter, the problem of elastoplastic modeling of an orthotropic boron-aluminum fiber-reinforced composite thick-walled rotating cylinder subjected to a temperature gradient has been discussed by using Seth's transition and generalized strain measure theory.

10.2 MATHEMATICAL MODELING AND GOVERNING EQUATION

10.2.1 MATHEMATICAL MODEL

Consider a thick-walled cylinder made of boron-aluminum fiber-reinforced composite material with internal and external radii 'a' and 'b,' respectively. The cylinder is rotating about its axis with an angular velocity ω and subjected to a steady-state temperature φ_0 on the inner surface of the cylinder.

10.2.2 FORMULATION OF THE PROBLEM

The components of displacement in cylindrical polar coordinates are given by

$$u = r(1-\beta), v = 0, w = dz \tag{2}$$

where, β is a function of r only and $r = \sqrt{x^2 + y^2}$, d is a constant which is the allowance for a uniform longitudinal extension.

The finite strain components are given by:

$$e_{rr}^A = \frac{\partial u}{\partial r} - \frac{1}{2}\left[\left(\frac{\partial u}{\partial r}\right)^2 + \left(\frac{\partial v}{\partial r}\right)^2 + \left(\frac{\partial w}{\partial r}\right)^2 - v^2\right] \tag{3}$$

$$= \frac{1}{2}[1-(\beta + r\beta')^2]$$

$$e_{\theta\theta}^A = \frac{1}{r}\frac{\partial v}{\partial \theta} + \frac{u}{r} - \frac{1}{2r^2}\left[\left(\frac{\partial u}{\partial \theta}\right)^2 + r^2\left(\frac{\partial v}{\partial \theta}\right)^2 + \left(\frac{\partial w}{\partial \theta}\right)^2\right] \tag{4}$$

$$-\frac{1}{2r^2}\left[-vr^2\frac{\partial u}{\partial \theta} + u\frac{\partial v}{\partial \theta} - v\frac{\partial u}{\partial \theta} + ur^2\frac{\partial v}{\partial \theta} + u^2 + r^2v^2\right]$$

$$= \frac{1}{2}[1-\beta^2]$$

$$e_{zz}^A = \frac{\partial w}{\partial z} - \frac{1}{2}\left[\left(\frac{\partial u}{\partial z}\right)^2 + r\left(\frac{\partial v}{\partial z}\right)^2 + \left(\frac{\partial w}{\partial z}\right)^2\right] \tag{5}$$

$$= \frac{1}{2}[1-(1-d)^2]$$

$$e_{r\theta}^A = e_{\theta z}^A = e_{zr}^A = 0$$

where, u, v, w are the physical components of displacement u_i and e_{rr}^A, $e_{\theta\theta}^A$, e_{zz}^A, $e_{r\theta}^A e_{\theta z}^A$, e_{zr}^A are the components of the strain tensor e_{ij}^A, the meaning of superscripts "A" is Almansi.

Using equations (3), (4), and (5) into equation (1), the generalized components of strain are given by:

$$e_{rr} = \frac{1}{n}[1-(\beta + r\beta')^n] \tag{6}$$

$$e_{\theta\theta} = \frac{1}{n}[1-\beta^n] \tag{7}$$

$$e_{zz} = \frac{1}{n}[1-(1-d)^n] \tag{8}$$

$$e_{r\theta} = e_{\theta z} = e_{zr} = 0$$

The stress-strain relations for an orthotropic material are given by Love [1],

$$\begin{bmatrix} \sigma_{rr} \\ \sigma_{\theta\theta} \\ \sigma_{zz} \\ \sigma_{\theta z} \\ \sigma_{zr} \\ \sigma_{r\theta} \end{bmatrix} = \begin{bmatrix} C_{11} & C_{12} & C_{13} & 0 & 0 & 0 \\ C_{21} & C_{22} & C_{23} & 0 & 0 & 0 \\ C_{31} & C_{32} & C_{33} & 0 & 0 & 0 \\ 0 & 0 & 0 & C_{44} & 0 & 0 \\ 0 & 0 & 0 & 0 & C_{55} & 0 \\ 0 & 0 & 0 & 0 & 0 & C_{66} \end{bmatrix} \begin{bmatrix} e_{rr} \\ e_{\theta\theta} \\ e_{zz} \\ e_{\theta z} \\ e_{zr} \\ e_{r\theta} \end{bmatrix} \tag{9}$$

where, C_{11}, C_{12}, C_{13}, C_{21}, C_{22}, C_{23}, C_{31}, C_{32}, C_{33}, C_{44}, C_{55} and C_{66} are the constants of the material.

Using equations (6), (7), and (8) in equation (9), the thermo-elastic constitutive equations are given by:

$$\sigma_{rr} = \frac{C_{11}}{n}\left[1-(\beta + r\beta')^n\right] + \frac{C_{12}}{n}\left[1-\beta^n\right] + \frac{C_{13}}{n}\left[1-(1-d)^n\right] - \alpha_1\varphi \tag{10}$$

$$\sigma_{\theta\theta} = \frac{C_{21}}{n}\left[1-(\beta + r\beta')^n\right] + \frac{C_{22}}{n}\left[1-\beta^n\right] + \frac{C_{23}}{n}\left[1-(1-d)^n\right] - \alpha_2\varphi \tag{11}$$

$$\sigma_{zz} = \frac{C_{31}}{n}\left[1-(\beta + r\beta')^n\right] + \frac{C_{32}}{n}\left[1-\beta^n\right] + \frac{C_{33}}{n}\left[1-(1-d)^n\right] - \alpha_3\varphi \tag{12}$$

$$e_{r\theta} = e_{\theta z} = e_{zr} = 0$$

where, α_i $(i = 1,2,3)$ being the coefficient of thermal expansion and φ is the temperature change.

The temperature field satisfying Fourier heat equation $\nabla^2 \varphi = 0$ and

$$\varphi = \varphi_o \text{ at } r = a$$
$$\varphi = 0 \text{ at } r = b$$

where, φ_o is a constant, given by Seth. Solving the heat equation using the given condition, one gets:

$$\varphi = \bar{\varphi}_o \ln(r/b) \tag{13}$$

where, $\bar{\varphi}_o = \dfrac{\varphi_o}{\ln(a/b)}$

The equations of equilibrium are all satisfied except:

$$\frac{d(r\sigma_{rr})}{dr} - \sigma_{\theta\theta} + \rho\omega^2 r^2 = 0 \tag{14}$$

where, ρ is the density of the material.

Substituting equations (10), (11) and (13) in equation (14), we get a non-linear differential equation in β as:

$$\Bigg[p(p+1)^n + \frac{C_{12}}{C_{11}} p + \frac{\alpha_1 \bar{\theta}_o}{C_{11}\beta^n} - \frac{1}{nC_{11}\beta^n}\{(C_{11} - C_{21})(1 - \beta^n (p+1)^n)$$
$$+ (C_{12} - C_{22})(1 - \beta^n) + (C_{13} - C_{23})(1 - (1-d)^n)$$
$$+ (\alpha_2 - \alpha_1) n\varphi\} - \frac{\rho\omega^2 r^2}{C_{11}\beta^n}\Bigg]\frac{d\beta}{dp} + \beta p (p+1)^{n-1} = 0 \tag{15}$$

10.2.3 BOUNDARY CONDITIONS

The boundary condition requires that

$$\sigma_{rr} = 0 \text{ at } r = a \text{ and } r = b \tag{16}$$

The resultant force normal to the plane z=constant must vanish, that is,

$$\int_a^b r\sigma_{zz}\, dr = 0 \tag{17}$$

10.3 SOLUTION OF THE PROBLEM

The asymptotic solution through the principal stress leads from elastic to plastic state at the transition point $p \to \pm\infty$. We define the transition function ξ as,

$$\xi = 1 - \frac{n}{C_{11}+C_{12}+C_{13}}\sigma_{rr} - \frac{(C_{11}-C_{21})\alpha_1\varphi n}{C_{11}(C_{11}+C_{12}+C_{13})} - \frac{n\rho\omega^2 r^2}{2(C_{11}+C_{12}+C_{13})} \tag{18}$$

Substituting equation (10) in equation (18), we get:

$$\xi = \frac{C_{11}}{C_{11}+C_{12}+C_{13}}\left[\beta^n(p+1)^n + \frac{C_{12}\beta^n}{C_{11}} + \frac{C_{13}(1-d)^n}{C_{11}} + \frac{\alpha_1 n\varphi}{C_{11}}\left(1-\frac{C_{11}-C_{21}}{C_{11}}\right) - \frac{n\rho\omega^2 r^2}{2}\right] \tag{19}$$

Then:

$$\frac{d(\ln\xi)}{dr} = \frac{n\beta^n C_{11}}{Rr(C_{11}+C_{12}+C_{13})}\left[p(p+1)^{n-1}\left((p+1)+\beta\frac{dp}{d\beta}\right) + \frac{C_{12}}{C_{11}}p + \frac{\alpha_1\bar{\varphi}_o}{C_{11}\beta^n}\left(1-\frac{C_{11}-C_{21}}{C_{11}}\right) - \frac{\rho\omega^2 r^2}{C_{11}\beta^n}\right] \tag{20}$$

Substituting the value of $\frac{dp}{d\beta}$ from equation (15) in equation (20), we get:

$$\frac{d(\ln\xi)}{dr} = \frac{n\beta^n C_{11}}{Rr(C_{11}+C_{12}+C_{13})}\left[\frac{\alpha_1\varphi_o}{C_{11}\beta^n} + \frac{1}{nC_{11}\beta^n}\left((C_{11}-C_{21})(1-\beta^n(p+1)^n) + (C_{12}-C_{22})(1-\beta^n) + (C_{13}-C_{23})(1-(1-d)^n)\right) + (\alpha_2-\alpha_1)n\bar{\varphi}_o\ln(r/b)) - \frac{\alpha_1\bar{\varphi}_o}{C_{11}\beta^n}\left(1-\frac{C_{11}-C_{21}}{C_{11}}\right)\right] \tag{21}$$

Taking the asymptotic value of equation (21) as $p \to \pm\infty$, we get:

$$\frac{d(\ln\xi)}{dr} = \frac{-(C_{11}-C_{21})}{C_{11}r} \tag{22}$$

Then from Eqn. (22)

$$\xi = A_1 r^{\frac{-(C_{11}-C_{21})}{C_{11}}} \tag{23}$$

where, A_1 is a constant of integration:

Using Eqn. (23) in Eqn. (18), we get:

$$\sigma_{rr} = \left(\frac{C_{11}+C_{12}+C_{13}}{n}\right)\left[1-A_1 r^{\frac{-(C_{11}-C_{21})}{C_{11}}}\right] - \frac{(C_{11}-C_{21})}{C_{11}}\alpha_1\varphi - \frac{\rho\omega^2 r^2}{2}$$

$$= \frac{\left(C_{22}C_{33}-C_{23}^2\right)\left(C_{11}+C_{12}+C_{13}\right)Y\left[1-A_1 r^{\frac{-(C_{11}-C_{21})}{C_{11}}}\right]}{[C_{11}C_{22}C_{33}-C_{23}^2 C_{11}-C_{33}C_{12}^2+2C_{12}^2 C_{23}-C_{13}^2 C_{22}]}$$

$$-\frac{(C_{11}-C_{21})}{C_{11}}\alpha_1\bar{\varphi}_o \ln(r/b) - \frac{\rho\omega^2 r^2}{2} \quad (24)$$

where, $y = \frac{\left[C_{11}C_{22}C_{33}-C_{23}^2 C_{11}-C_{33}C_{12}^2+2C_{12}^2 C_{23}-C_{13}^2 C_{22}\right]}{\left[C_{22}C_{33}-C_{23}^2\right]n}$, the yield stress in tension at the transition given by Seth.

Substituting Eqns. (13) and (24) in Eqn. (14), we get

$$\sigma_{\theta\theta} = \frac{\left(C_{22}C_{33}-C_{23}^2\right)\left(C_{11}+C_{12}+C_{13}\right)y\left[1-\left(1-\frac{C_{11}-C_{21}}{C_{11}}\right)A_1 r^{\frac{-(C_{11}-C_{21})}{C_{11}}}\right]}{[C_{11}C_{22}C_{33}-C_{23}^2 C_{11}-C_{33}C_{12}^2+2C_{12}^2 C_{23}-C_{13}^2 C_{22}]}$$

$$-\frac{(C_{11}-C_{21})}{C_{11}}\alpha_1\bar{\varphi}_o\left[\ln\left(\frac{r}{b}\right)+1\right] - \frac{\rho\omega^2 r^2}{2} \quad (25)$$

Equation (12) becomes

$$\sigma_{zz} = \frac{C_{31}}{C_{11}+C_{21}}[\sigma_{rr}+\sigma_{\theta\theta}] + \varphi\left[\frac{C_{31}}{C_{11}+C_{21}}(\alpha_1+\alpha_2)-\alpha_3\right]+2A_2 \quad (26)$$

where

$$A_2 = \frac{1}{2}\left[\frac{1}{n}\left(C_{32}-\frac{C_{31}(C_{12}+C_{22})}{(C_{11}+C_{21})}\right)+\left(C_{33}-\frac{C_{31}(C_{13}+C_{23})}{(C_{11}+C_{21})}\right)\left(1-(1-d)^n\right)\frac{1}{n}\right]$$

Let $D_1 = \left[C_{33}C_{22}-C_{23}^2\right]\left[C_{11}+C_{13}+C_{12}\right]$ and

$$D_2 = \left[C_{11}C_{22}C_{33}-C_{23}^2 C_{11}-C_{33}C_{12}^2+2C_{12}^2 C_{23}-C_{13}^2 C_{22}\right]$$

Using the boundary condition $\sigma_{rr} = 0$ at $r = a$ in Eqn. (24), we get:

$$A_1 = a^{\frac{C_{11}-C_{21}}{C_{11}}}\left[1-\frac{D_2}{D_1 y}\left(\frac{C_{11}-C_{21}}{C_{11}}\right)\alpha_1\varphi_o - \frac{D_2}{D_1 y}\frac{\rho\omega^2 a^2}{2}\right] \tag{27}$$

Using the boundary condition $\sigma_{rr} = 0$ at $r = b$ in Eqn. (24), we get:

$$A_1 = b^{\frac{C_{11}-C_{21}}{C_{11}}}\left[1-\frac{D_2}{D_1 y}\frac{\rho\omega^2 a^2}{2}\right] \tag{28}$$

Using Eqn. (27) and Eqn. (28), we get:

$$\frac{\rho\omega^2}{2} = \frac{\left(\frac{C_{11}-C_{21}}{C_{11}}\right)\alpha_1\varphi_o\left(\frac{b}{a}\right)^{-\left(\frac{C_{11}-C_{21}}{C_{11}}\right)} - \frac{D_1 y}{D_2}\left(\left(\frac{b}{a}\right)^{-\left(\frac{C_{11}-C_{21}}{C_{11}}\right)}-1\right)}{b^2 - a^2\left(\frac{b}{a}\right)^{-\left(\frac{C_{11}-C_{21}}{C_{11}}\right)}} \tag{29}$$

Using the boundary condition $\int_a^b r\sigma_{zz}\,dr = 0$ in Eqn. (26), we get:

$$A_2 = \frac{-C_{31}}{C_{11}+C_{21}}\frac{\rho\omega^2}{4}\left(a^2+b^2\right) - \bar{\varphi}_o\left[(\alpha_1+\alpha_2)\frac{C_{31}}{C_{11}+C_{21}} - \alpha_3\right]$$
$$\left[\frac{a^2}{4} - \frac{b^2}{4} - \frac{a^2}{2}\ln(a/b)\right] \tag{30}$$

Using the value of A_1 and A_2 from Eqns. (27) and (30) in Eqns. (24), (25) and (26), we get:

$$\sigma_{rr} = \frac{D_1 y}{D_2}\left[1-\left(\frac{a}{r}\right)^{\frac{C_{11}-C_{21}}{C_{11}}}\left[1-\frac{D_2}{D_1 y}\left(\frac{C_{11}-C_{21}}{C_{11}}\right)\alpha_1\varphi_o - \frac{D_2}{D_1 y}\frac{\rho\omega^2 a^2}{2}\right]\right]$$
$$-\frac{(C_{11}-C_{21})}{C_{11}}\alpha_1\bar{\varphi}_o\ln(r/b) - \frac{\rho\omega^2 r^2}{2} \tag{31}$$

$$\sigma_{\theta\theta} = \frac{D_1 y}{D_2}\left[1-\left(1-\frac{C_{11}-C_{21}}{C_{11}}\right)\left(\frac{a}{r}\right)^{\frac{C_{11}-C_{21}}{C_{11}}}\left[1-\frac{D_2}{D_1 y}\left(\frac{C_{11}-C_{21}}{C_{11}}\right)\alpha_1\varphi_o - \frac{D_2}{D_1}\frac{\rho\omega^2 a^2}{2}\right]\right] -$$

$$\frac{(C_{11}-C_{21})}{C_{11}}\alpha_1\bar{\varphi}_o\left(\ln\left(\frac{r}{b}\right)+1\right) - \frac{\rho\omega^2 r^2}{2} \tag{32}$$

$$\sigma_{zz} = \frac{C_{31}}{C_{11}+C_{21}}[\sigma_{rr}+\sigma_{\theta\theta}]+\bar{\varphi}_o \ln\left(\frac{r}{b}\right)\left[\frac{C_{31}}{C_{11}+C_{21}}(\alpha_1+\alpha_2)-\alpha_3\right]-\frac{2\bar{\varphi}_o}{b^2-a^2}$$

$$\left[\frac{C_{31}}{C_{11}+C_{21}}(\alpha_1+\alpha_2)-\alpha_3\right]\left[\frac{a^2}{4}-\frac{b^2}{4}-\frac{a^2}{2}\ln(a/b)\right]-\frac{C_{31}}{C_{11}+C_{21}}\frac{\rho\omega^2}{2}(a^2+b^2) \quad (33)$$

10.3.1 INITIAL YIELDING

It is found that the value of $|\sigma_{rr}-\sigma_{\theta\theta}|$ is maximum at $r=a$, which means that yielding will take place at the inner surface of the cylinder for orthotropic material. Thus:

$$|\sigma_{rr}-\sigma_{\theta\theta}| = \frac{D_1 y}{D_2}.\frac{C_{11}-C_{21}}{C_{11}}\left[1-\frac{D_2}{D_1 y}\left(\frac{C_{11}-C_{21}}{C_{11}}\right)\alpha_1\varphi_o-\frac{D_2}{D_1 y}\frac{\rho\omega^2 a^2}{2}\right]$$

$$-\frac{C_{11}-C_{21}}{C_{11}}\alpha_1\bar{\varphi}_o = Y \quad (34)$$

Using Eqn. (34)

$$y = \frac{D_2}{D_1}\left(\frac{C_{11}}{C_{11}-C_{21}}Y\right)+\frac{D_2}{D_1}\left(\frac{C_{11}-C_{21}}{C_{11}}\right)\alpha_1\varphi_o+\frac{D_2}{D_1}\frac{\rho\omega^2 a^2}{2}+\frac{D_2}{D_1}\alpha_1\bar{\varphi}_o \quad (35)$$

Substituting Eqn. (35) in Eqns. (31), (32) and (33), we get:

$$\sigma_{rr} = \frac{C_{11}Y}{C_{11}-C_{21}}\left(1-\left(\frac{a}{r}\right)^{\frac{C_{11}-C_{21}}{C_{11}}}\right)+\beta_o\left[\frac{1-\left(\frac{a}{r}\right)^{\frac{C_{11}-C_{21}}{C_{11}}}+\left(\frac{C_{11}-C_{21}}{C_{11}}\right)\ln(a/r)}{\ln(a/b)}\right]$$

$$+\frac{\rho\omega^2}{2}(a^2-r^2) \quad (36)$$

$$\sigma_{\theta\theta} = \frac{C_{11}Y}{C_{11}-C_{21}}\left(1-\left(1-\frac{C_{11}-C_{21}}{C_{11}}\right)\left(\frac{a}{r}\right)^{\frac{C_{11}-C_{21}}{C_{11}}}\right)+\beta_o$$

$$\left[\frac{1-\left(1-\frac{C_{11}-C_{21}}{C_{11}}\right)\left(\frac{a}{r}\right)^{\frac{C_{11}-C_{21}}{C_{11}}}-\left(\frac{C_{11}-C_{21}}{C_{11}}\right)\left(\ln\left(\frac{a}{r}\right)-1\right)}{\ln(a/b)}\right]+\frac{\rho\omega^2}{2}(a^2-r^2) \quad (37)$$

$$\sigma_{zz}=\frac{C_{31}}{C_{11}+C_{21}}[\sigma_{rr}+\sigma_{\theta\theta}]+\left[\frac{C_{31}}{C_{11}+C_{21}}(\alpha_1+\alpha_2)-\alpha_3\right]$$
$$\left[\bar{\varphi}_o\ln\left(\frac{r}{b}\right)-\frac{2\bar{\varphi}_o}{b^2-a^2}\left(\frac{a^2}{4}-\frac{b^2}{4}-\frac{a^2}{2}\ln(a/b)\right)\right]-\frac{C_{31}}{C_{11}+C_{21}}\frac{\rho\omega^2}{2}(a^2+b^2) \quad (38)$$

$$\frac{\rho\omega^2}{2}=\frac{\left(\frac{C_{11}-C_{21}}{C_{11}}\right)\alpha_1\varphi_o\left(\frac{b}{a}\right)^{-\left(\frac{C_{11}-C_{21}}{C_{11}}\right)}-\left(\left(\frac{b}{a}\right)^{-\left(\frac{C_{11}-C_{21}}{C_{11}}\right)}-1\right)\left(\frac{C_{11}}{C_{11}-C_{21}}Y+\frac{C_{11}-C_{21}}{C_{11}}\alpha_1\varphi_o+\frac{\rho\omega^2a^2}{2}+\alpha_1\bar{\varphi}_o\right)}{b^2-a^2\left(\frac{b}{a}\right)^{-\left(\frac{C_{11}-C_{21}}{C_{11}}\right)}} \quad (39)$$

where, $\beta_o=\alpha_1\varphi_o$

For fully plastic,

$$C_{12}=C_{13}=C_{11},\ C_{21}=C_{23}=C_{22}\text{ and }C_{31}=C_{32}=C_{33}$$

Thus, Eqn. (39) becomes:

$$\sigma_{rr}=\frac{C_{11}Y}{C_{11}-C_{22}}\left(1-\left(\frac{a}{r}\right)^{\frac{C_{11}-C_{22}}{C_{11}}}\right)+\beta_o\left[\frac{1-\left(\frac{a}{r}\right)^{\frac{C_{11}-C_{22}}{C_{11}}}+\left(\frac{C_{11}-C_{22}}{C_{11}}\right)\ln(a/r)}{\ln(a/b)}\right]$$
$$+\frac{\rho\omega^2}{2}(a^2-r^2) \quad (40)$$

$$\sigma_{\theta\theta}=\frac{C_{11}Y}{C_{11}-C_{22}}\left(1-\left(1-\frac{C_{11}-C_{22}}{C_{11}}\right)\left(\frac{a}{r}\right)^{\frac{C_{11}-C_{22}}{C_{11}}}\right)+\beta_o$$
$$\left[\frac{1-\left(1-\frac{C_{11}-C_{22}}{C_{11}}\right)\left(\frac{a}{r}\right)^{\frac{C_{11}-C_{22}}{C_{11}}}+\left(\frac{C_{11}-C_{22}}{C_{11}}\right)\left(\ln\left(\frac{a}{r}\right)-1\right)}{\ln(a/b)}\right]+\frac{\rho\omega^2}{2}(a^2-r^2) \quad (41)$$

$$\sigma_{zz}=\frac{C_{33}}{C_{11}+C_{22}}[\sigma_{rr}+\sigma_{\theta\theta}]+\left[\frac{C_{33}}{C_{11}+C_{22}}(\alpha_1+\alpha_2)-\alpha_3\right]$$
$$\left[\bar{\varphi}_o\ln\left(\frac{r}{b}\right)-\frac{2\bar{\varphi}_o}{b^2-a^2}\left(\frac{a^2}{4}-\frac{b^2}{4}-\frac{a^2}{2}\ln(a/b)\right)\right]-\frac{C_{33}}{C_{11}+C_{22}}\frac{\rho\omega^2}{2}(a^2+b^2) \quad (42)$$

$$\frac{\rho\omega^2}{2}=\frac{\left(\frac{C_{11}-C_{22}}{C_{11}}\right)\alpha_1\varphi_o\left(\frac{b}{a}\right)^{-\left(\frac{C_{11}-C_{22}}{C_{11}}\right)}-\left(\left(\frac{b}{a}\right)^{-\left(\frac{C_{11}-C_{22}}{C_{11}}\right)}-1\right)\left(\frac{C_{11}}{C_{11}-C_{22}}Y+\frac{C_{11}-C_{22}}{C_{11}}\alpha_1\varphi_o+\frac{\rho\omega^2a^2}{2}+\alpha_1\bar{\varphi}_o\right)}{b^2-a^2\left(\frac{b}{a}\right)^{-\left(\frac{C_{11}-C_{22}}{C_{11}}\right)}} \tag{43}$$

We introduce the following non-dimensional quantities:

$$R_o=\frac{b}{a},R=\frac{a}{r},\sigma_r=\frac{\sigma_{rr}}{Y},\sigma_\theta=\frac{\sigma_{\theta\theta}}{Y},\Omega^2=\frac{\rho\omega^2a^2}{2Y}$$

Elastoplastic stresses and angular speed from Eqns. (36), (37), (38), and (39) in non-dimensional form become:

$$\sigma_r=\frac{C_{11}}{C_{11}-C_{21}}\left(1-R^{\frac{C_{11}-C_{21}}{C_{11}}}\right)+\frac{\beta_o}{Y}\left[\frac{1-R^{\frac{C_{11}-C_{21}}{C_{11}}}+\left(\frac{C_{11}-C_{21}}{C_{11}}\right)\ln(R)}{-\ln(R_o)}\right]+\Omega^2\left(1-R^{-2}\right) \tag{44}$$

$$\sigma_\theta=\frac{C_{11}}{C_{11}-C_{21}}\left(1-\left(1-\frac{C_{11}-C_{21}}{C_{11}}\right)R^{\frac{C_{11}-C_{21}}{C_{11}}}\right)+\frac{\beta_o}{Y}$$
$$\left[\frac{1-\left(1-\frac{C_{11}-C_{21}}{C_{11}}\right)R^{\frac{C_{11}-C_{21}}{C_{11}}}+\left(\frac{C_{11}-C_{21}}{C_{11}}\right)\left(\ln(R)-1\right)}{-\ln(R_o)}\right]+\Omega^2\left(1-R^{-2}\right) \tag{45}$$

$$\sigma_z=\frac{C_{31}}{C_{11}+C_{21}}[\sigma_r+\sigma_\theta]+\frac{\beta_O}{Y}\frac{1}{(-\ln(R_o))}\left[\frac{C_{31}}{C_{11}+C_{21}}\left(1-\frac{\alpha_2}{\alpha_1}\right)-\frac{\alpha_3}{\alpha_1}\right]$$
$$\left[-\ln(RR_o)-\frac{2}{R_o^2-1}\left(\frac{1}{4}-\frac{R_o^2}{4}+\frac{1}{2}\ln(R_o)\right)\right]-\frac{C_{31}}{C_{11}+C_{21}}\Omega^2(1+R_o^2) \tag{46}$$

$$\Omega^2=\frac{\frac{\beta_o}{Y}\left(\frac{C_{11}-C_{21}}{C_{11}}\right)R_o^{-\left(\frac{C_{11}-C_{21}}{C_{11}}\right)}-\left(R_o^{-\left(\frac{C_{11}-C_{21}}{C_{11}}\right)}-1\right)\left(\frac{C_{11}}{C_{11}-C_{21}}+\frac{C_{11}-C_{21}}{C_{11}}\frac{\beta_O}{Y}+\Omega^2+\frac{\beta_O}{Y}\frac{1}{(-\ln(R_o))}\right)}{R_O^2-R_o^{\left(\frac{C_{11}-C_{21}}{C_{11}}\right)}} \tag{47}$$

10.4 NUMERICAL ILLUSTRATION AND DISCUSSION

In Table 10.1, elastic constants C_{ij} for orthotropic material, boron-aluminum has been given [8]. In Figures 10.1–10.4, curves for the rotating cylinder made of boron aluminum material have been drawn for the transitional and fully plastic state.

In Figures 10.1–10.4 we use R_o instead of R_o, βo/Y instead of $\frac{\beta_o}{Y}$ and Ω^2 instead of Ω^2.

TABLE 10.1 Elastic Stiffness Constants of a 48 vol % Boron-Aluminum Fiber-Reinforced Composite, in Units of GPa

Material	C_{11}	C_{22}	C_{33}	C_{44}	C_{55}	C_{66}	C_{12}	C_{13}	C_{23}
Boron-Aluminum	184	181.7	243.6	54.5	55.2	50.3	74.2	59.7	58.8

It is observed from Figure 10.1(a) that a cylinder having a smaller radii ratio has higher angular speed at any values of the temperature as compared to a cylinder having a higher radii ratio for the transition state. Further, the angular speed of the cylinder increased as the temperature of the cylinder increases for each radii ratio of the cylinder. In Figure 10.1(b), the curves are drawn between temperature and angular speed for various thickness ratios for a fully plastic state. It is seen that cylinders have a higher angular speed for smaller radii ratios than cylinders that have a higher radii ratio at any values of the temperature. But for each radii ratio, the value of the angular speed is constant as the temperature increases.

It is seen from Figure 10.2 that cylinders having smaller radii ratios require higher angular speed for yielding as compared to cylinders having higher radii ratios. With the inclusion of thermal effects, the angular speed increased for initial yielding to a smaller radii ratio but for the fully plastic state, the angular speed is the same.

In Figure 10.3, the curves are drawn for stresses along with the radii ratio for the parameters of temperature and angular speed taken from Figure 10.1 (a) from the transitional state. It is observed that the maximum magnitude of the axial stress occurs at the external surface at high temperature for the transitional state as compared to circumferential and radial stresses whereas the circumferential stress is maximum at the internal surface at any temperature and angular speed. The value of axial stress at low temperature and angular speed is between radial and circumferential stresses but as the temperature and angular speed increase the axial stress decreased and it is below both radial and circumferential stresses.

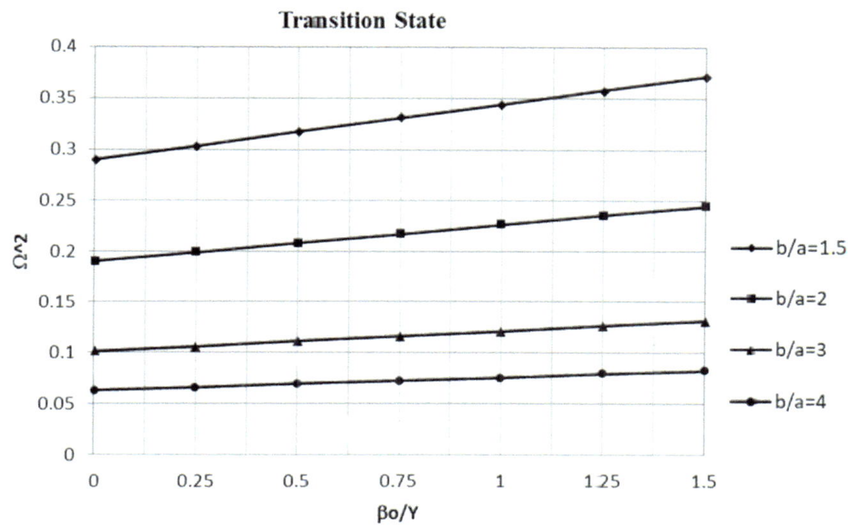

(a)

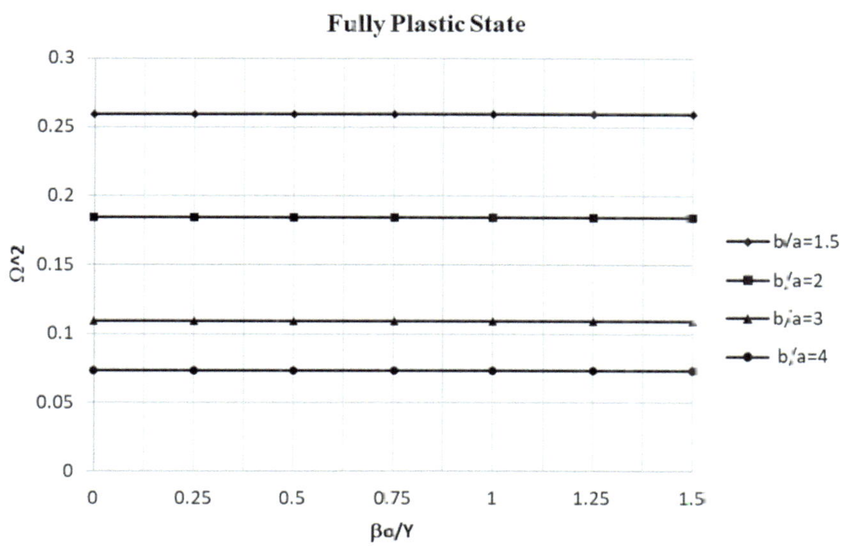

(b)

FIGURE 10.1 Relation between Ω^2 and βo/Y under different thickness ratios.

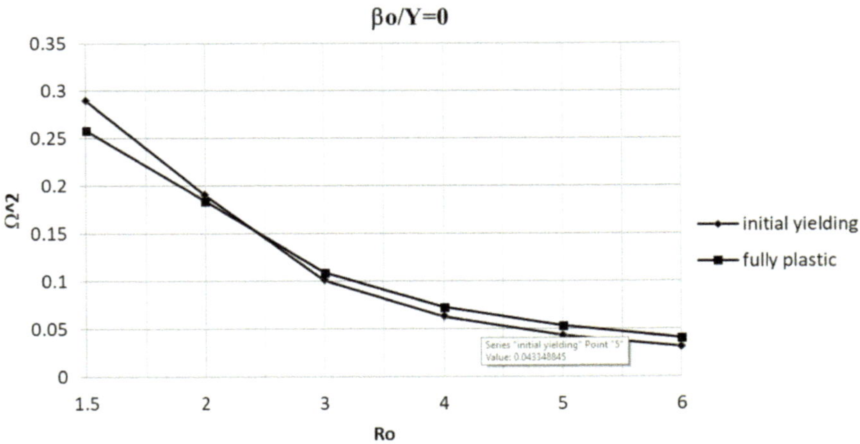

(a)

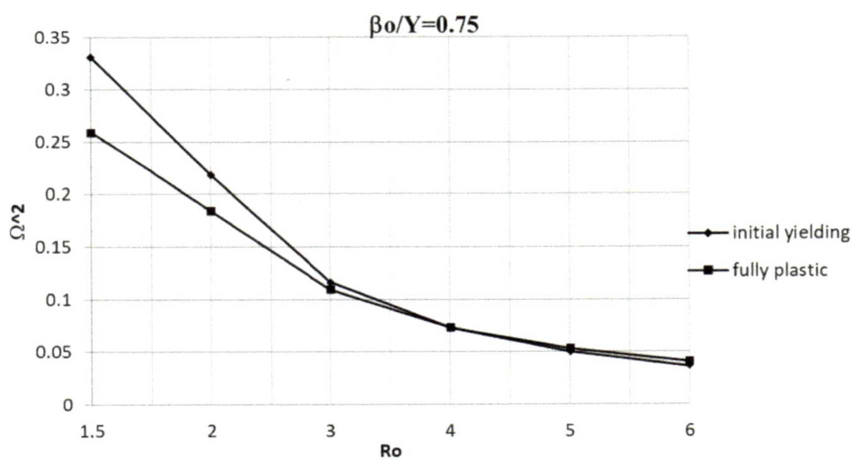

(b)

FIGURE 10.2 Angular speed required for transitional and fully plastic state.

It is seen from Figure 10.4 that the maximum circumferential stress occurs at the internal surface for the fully plastic state at any temperature and angular speed and it is also seen that the axial stress increased as both the temperature and angular speed increase.

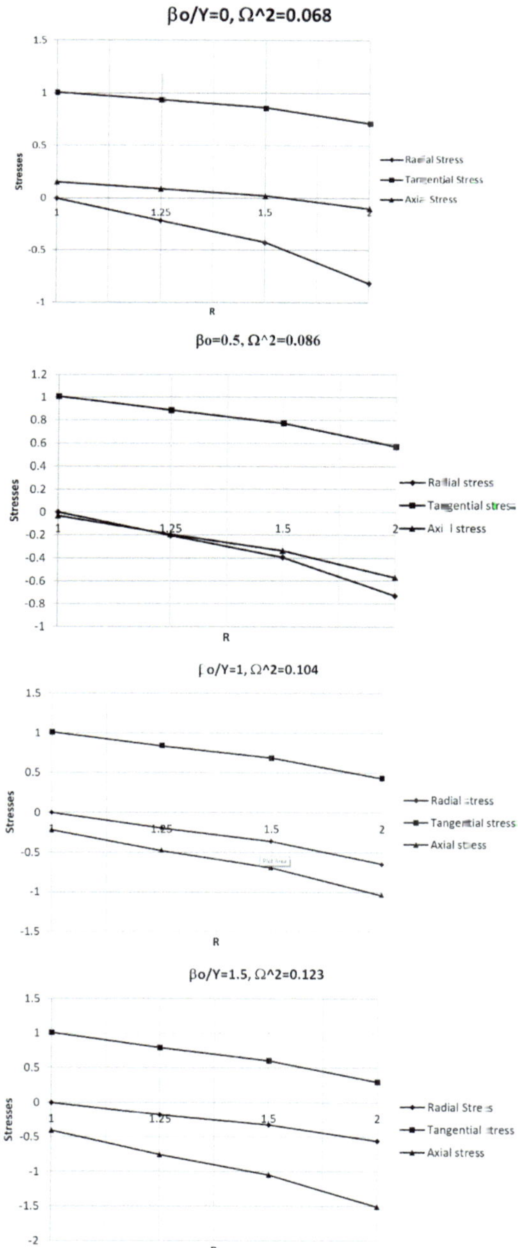

(a)

(b)

(c)

(d)

FIGURE 10.3 Distribution of orthotropic transitional plastic stresses due to rotation and temperature through the wall of the cylinder.

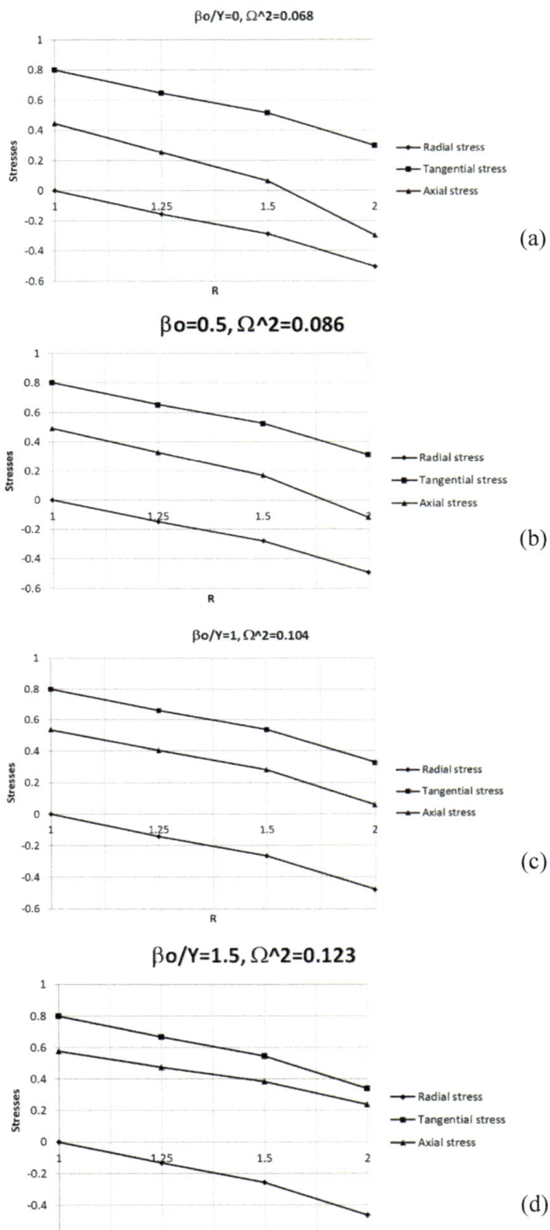

FIGURE 10.4 Distribution of boron-aluminum fully plastic stresses due to rotation and temperature through the wall of the cylinder.

10.5 CONCLUSION

It has been observed that a cylinder having a smaller radii ratio has higher angular speed at any values of temperature as compared to a cylinder having a higher radii ratio for both transition and fully plastic state. It is also seen that the angular speed of the cylinder increased for the transition state whereas for the fully plastic state the value of the angular speed is constant as the temperature of the cylinder increases for each radii ratio. It is seen that cylinders having smaller radii ratios require higher angular speed for yielding as compared to cylinders having higher radii ratios. With the inclusion of thermal effects, the angular speed increased for initial yielding to a smaller radii ratio but for the fully plastic state, the angular speed is the same. It is observed that the maximum circumferential stress occurs at the internal surface for both transitional and fully plastic state at any temperature and angular speed but the axial stress decreased for the transitional state and increased for the fully plastic state as both the temperature and angular speed increases.

KEYWORDS

- **elastoplastic**
- **orthotropic material**
- **rotating cylinder**
- **strain**
- **stress**
- **temperature gradient**

REFERENCES

1. Love, A. E. H., & M. A., (1944). *A Treatise on the Mathematical theory of Elasticity* (4th edn.).
2. Temesgen, A., Singh, S. B., & Pankaj, T., (2019). *Classical and Non-classical Treatment of Problems in Elastic-Plastic and Creep Deformation for Rotating Discs*. Canada: Apple Academic Press.

3. Davis, E. A., & Connelly, F. M., (1959). Stress distribution and plastic deformation in rotating cylinders strain hardening material. *Jr. Appl. Mech., 26*, 25–30.
4. Fung, Y. C., (1965). *Foundations of Solid Mechanics*. Engle wood Cliffs, NJ: Prentice-Hall.
5. Gupta, S. K., & Bhardwaj, P. C., (1986). Elastic-plastic and creep transition in an orthotropic rotating cylinder. *Proc. Indian Natn. Sci. Acad., 52A*(6), 1357–1369.
6. Gupta, S. K., & Rana, V. D., (1983). Thermo-elastic-plastic transition in rotating cylinders. *Indian Jr. of Tech., 21*, 499–502.
7. Gupta, S. K., & Shukla, R. K., (1991). Elastic-plastic transition in an orthotropic shell under internal pressure. *Indian J. Pure Appl. Math., 23*(31), 243–250.
8. Hassel, L., Christopher, F., & Paul, H., (1995). Orthotropic elastic constants of a boron-aluminum fiber-reinforced composite: An Acoustic resonance spectroscopy study. *Journal of Applied Physics, 78*, 1542.
9. Pankaj, T., (2009). Elastic-plastic transition stresses in a transversely isotropic thick-walled cylinder subjected to internal pressure and steady-state temperature. *Thermal Science, 13*(4), 107–118.
10. Seth, B. R., (1966). Measure concept in mechanics. *Int. J. Non-linear Mech., I*(2), 35–40.
11. Seth, B. R., (1964). On the problems of transition phenomenon. *Bull. Inst. Politec. Roum., 10*, 255–262.
12. Seth, B. R., (1970). Transition analysis of the collapse of thick-walled cylinders. *ZAMM, 50*, 617–621.
13. Shukla, R. K., (1997). Elastic-plastic transition in a compressible cylinder under internal pressure. *Indian J. Pure Appld. Math., 28*(2), 277–288.
14. Singh, S. B., & Ray, S., (2002). Modeling the anisotropy and creep in orthotropic aluminum-silicon carbide composite rotating disc. *Mechanics of Materials* (Vol. 34, No. 6, pp. 363–372). Elsevier Publishers.
15. Singh, S. B., & Ray, S., (2003). Newly proposed yield criterion for residual stress and steady-state creep in an anisotropic composite rotating disc. *Journal of Materials Processing Technology* (Vol. 143, No. 144C, pp. 623–628). Elsevier Publishers.
16. Sokolinikoff, I. S., (1950). *Mathematical Theory of Elasticity* (2nd edn., pp. 70–71). New York: McGraw-Hill Book Co.
17. Timoshenko, S. P., & Goodier, J. N., (1951). *Theory of Elasticity* (3rd edn.). New York, London: Mc Graw-Hall Book Co.

Index

T

U

V

W

X

Y

Z